Robert Blust
A Dictionary of Austronesian Monosyllabic Roots

Pacific Linguistics

Managing editor
Alexander Adelaar

Editorial board members
Wayan Arka
Danielle Barth
Don Daniels
T. Mark Ellison
Bethwyn Evans
Nicholas Evans
Gwendolyn Hyslop
David Nash
Bruno Olsson
Bill Palmer
Andrew Pawley
Malcolm Ross
Dineke Schokkin
Jane Simpson

Volume 652

Robert Blust

A Dictionary of Austronesian Monosyllabic Roots (Submorphemes)

—

ISBN 978-3-11-135778-2
e-ISBN (PDF) 978-3-11-078169-4
e-ISBN (EPUB) 978-3-11-078177-9
ISSN 1448-8310

Library of Congress Control Number: 2022931098

Bibliographic information published by the Deutsche Nationalbibliothek
The Deutsche Nationalbibliothek lists this publication in the Deutsche Nationalbibliografie;
detailed bibliographic data are available on the Internet at http://dnb.dnb.de.

© 2023 Walter de Gruyter GmbH, Berlin/Boston
This volume is text- and page-identical with the hardback published in 2022.
Photo credit: yuriz/iStock/Getty Images Plus
Typesetting: Integra Software Services Pvt. Ltd.
Printing and binding: CPI books GmbH, Leck

www.degruyter.com

To Charles Gober, 1945–1966,
Who gave his life to a country that did not love him,
From a friend who will not forget him.

Contents

List of abbreviations —— IX

Introduction —— 1

1 Literature review —— 4

2 Issues relating to the root in Austronesian languages —— 27
    2.1   Finding the root —— 28
    2.2   Root ambiguity —— 32
    2.3   Control of chance —— 34
    2.4   Etymological independence/control of duplication —— 36
    2.5   Root and doublet —— 39
    2.6   The shape/structure of the root —— 40
    2.7   Referents of roots —— 47
    2.8   The origin of roots —— 50
    2.9   False cognates? —— 50
    2.10   The problem of transmission —— 58

3 Summary of the data —— 60

4 Sources for the data —— 80

5 The Data —— 87

6 Appendices —— 265
    Appendix 1 —— 265
    Appendix 2 —— 268

References —— 271

List of abbreviations

ACD	Austronesian Comparative Dictionary (https://www.trussel2.com/acd , https://acd.clld.org/)
AN	Austronesian
BM	Bolaang Mongondow (Dunnebier 1951)
C	consonant
dbl.	doublet
cp.	compare
DPB	Dairi-Pakpak Batak (Manik 1977)
EIM	etymologically independent morpheme
k.o.	kind of
M	metathesis
MP	Malayo-Polynesian
N	nasal
n.d.	no date
OC	Order Class
onom.	onomatopoetic
opp.	opposite
o.s.	of something
PAN	Proto-Austronesian (Blust and Trussel 2020)
PBT	Proto-Bungku-Tolaki (Mead 1998)
PCMP	Proto-Central Malayo-Polynesian (Blust and Trussel 2020)
PCEMP	Proto-Central-Eastern Malayo-Polynesian (Blust and Trussel 2020)
PEG	Proto-East Gorontalic (Usup 1981)
PMic	Proto-Micronesian (Bender et al. 2003)
PMP	Proto-Malayo-Polynesian (Blust and Trussel 2020)
PNV	Proto-North Vanuatu (Clark 2009)
POC	Proto-Oceanic (Blust and Trussel 2020)
PPH	Proto-Philippines (Blust and Trussel 2020)
PR	Proto-Rukai (Li 1977)
PS	Proto-Sangiric (Sneddon 1984)
PSS	Proto-South Sulawesi (Mills 1975)
PWMP	Proto-Western-Malayo-Polynesian
s.o.	someone
s.t.	something
V	vowel
WBM	Manobo, Western Bukidnon (Elkins 1968)
WMP	Western-Malayo-Polynesian

Introduction

In an earlier publication (Blust 1988b) I attempted to revive interest in a long-neglected topic first addressed by Brandstetter in 1910, namely the frequent recurrence of semantically similar submorphemic -CVC sequences in Austronesian (AN) languages. Brandstetter called such morpheme-forming elements 'roots' (Wurzeln), and I followed his practice in my earlier treatment, and continue to do so here.

The root in AN languages can most usefully be compared to the phonestheme in English or other languages in which sound symbolism has been studied. In both cases a sound-meaning association is established by recurrence, but morphemes are additionally defined by contrast, while the root or phonestheme is not. In English this is seen clearly in cases like *cat*: *cat-s*, *frog*: *frog-s*, or *horse*: *horse-s* as against *gleam, glimmer, glitter, glisten, glow*, and the like. Both *-s* and *gl-* are recurrently associated with a meaning ('plural' and 'light, luminosity' respectively), but *-s* is separable from the stem with which it occurs (it is a morpheme), whereas *gl-* cannot be separated from the rest of the morpheme in which it occurs and leave a meaningful residue (it is a phonestheme). Like the phonestheme in English, then, the root in AN languages is a sound-meaning association that can be established by recurrence, but not by contrast. To illustrate, Proto-Malayo-Polynesian *delem 'dark of the moon, moonless night', *halem 'night; dark', Proto-Austronesian *Sulem 'dimness, twilight', Ngaju Dayak *bilem* 'black', Ilokano *lúlem* 'overcast, clouded over, darkish', and Kavalan *telem* 'dark' all share the monosyllabic root *-lem 'dark', but if this meaningful element is removed, the remainder is a meaningless phoneme string, just as the residue of words like *gleam, glimmer* and the like has no meaning in English without the initial consonant cluster.

Blust (1988b) established the existence of 231 roots in AN languages. The minimal criterion for accepting a recurrent submorphemic sound-meaning correlation as a root was the recognition of at least four etymologically independent morphemes (hereafter EIMs). In order not to artificially inflate the database, cognate morphemes that share the same root over a number of languages were treated as a single case represented by a reconstruction at the level of Proto-Austronesian (PAN), Proto-Malayo-Polynesian (PMP), or Proto-Western-Malayo-

Note: I am indebted to two anonymous reviewers and to April Almarines, Alex Smith, and R. David Zorc for their careful reading of the preliminary manuscript, which led to a number of improvements in the quality of the text. Any errors that remain are mine alone.

Polynesian (PWMP). Although they may be spread over many languages, then, cognate sets are treated in the same way as morphemes found in a single attested language or very low-level subgroup. This distinction among ancestral stages has been extended in the present work to include Proto-Philippines (PPH), Proto-Central-Eastern Malayo-Polynesian (PCEMP), and in rare cases other proto-languages. Although Western-Malayo-Polynesian is no longer considered a valid subgroup (Smith 2017), a WMP node is recognized as a default for what may be several different proto-languages that form primary branches of MP. Until more extensive work is done in definitively establishing smaller subgroups within MP, there appears to be no satisfactory alternative to this practice, which is continued here.

The 231 roots in Blust (1988b, Appendix 3) occur in 2,459 EIMs (reconstructed forms, and forms currently known only in single languages or very close relatives). These range from 48 EIMs with *-pit 'press, squeeze together; narrow' (hence 'to approach one another, of two surfaces') to 4 EIMs for fifteen roots that are ordered alphabetically from *-dek 'pulverized, pounded fine' to *-tir 'to tremble, quiver' (now revised to *-tiR). A number of issues connected with the establishment of roots and their properties were discussed in that publication, and some of these will be reviewed again here.

During the more than three decades between Blust (1988b) and the writing of the present work, the *Austronesian Comparative Dictionary* (ACD) came into being and grew considerably in scope (Blust and Trussel 2020). As part of that large project, a separate file of monosyllabic roots was compiled whenever these were identified in reconstructed morphemes. This greatly expanded the earlier corpus, and when Steve Trussel suddenly passed away on June 27, 2020, these were stranded along with the rest of the dictionary in a kind of suspended animation.

Recently it occurred to me that the collection of roots in the ACD could be organized as a separate dictionary, and expanded through continued searching. The present work, which includes 406 roots supported by 5,404 EIMs, represents an increase of over 75% in number of roots identified, and more than double the number of morphemes that exemplify them. Although it inevitably remains unfinished, and probably always will be, the corpus of material presented here is easily the largest, and methodologically the most controlled collection of widespread submorphemes in AN languages.

Apart from the data itself the remainder of this book will address the following topics: 1. a literature review, and 2. a survey of issues relating to the -CVC root in AN languages.

This second topic is subdivided into 2.1. Finding the root, 2.2. Root ambiguity, 2.3. Control of chance, 2.4. Etymological independence, 2.5. Root and doublet, 2.6. The shape of the root, 2.7. Referents of roots, 2.8. False cognates?, and 2.9. The problem of transmission.

1 Literature review

When Blust (1988b) appeared, little attention had been given to the question of the root in AN languages since the pioneering work of Brandstetter (1910). It is true (as pointed out by McCune 1985, Nothofer 1990, 1991, and some other writers) that a few scattered remarks on sound symbolism in Malay appear in the work of the British colonial officials C.N. Maxwell (1936), and R.J. Wilkinson (1936), and of the Indonesian scholar S.T. Alisjahbana (1949–50), as well as some passing remarks of questionable value on the -CVC root in Chamorro in Costenoble (1940), and that similar sporadic comments by W.E. Maxwell (1882) even preceded Brandstetter, but judged by their impact on scholarship in the field of AN linguistics, none of these was of major importance. Brandstetter's work was carefully reviewed in my monograph, and an examination of the strengths and weaknesses of his approach need not be repeated here. However, the appearance of my earlier book and the conference paper that preceded it (Blust 1988a) triggered a burst of interest in this question which rapidly resulted in several publications of varying quality. The first of these apparently was McCune (1985)[1].

McCune (1985). While Brandstetter (1916) used 'Indonesian' as a collective term that refers to all non-Oceanic AN languages, McCune uses the same term to describe a single language, namely Bahasa Indonesia, the national language of the Republic of Indonesia. This limitation to a single language for a phenomenon that is widespread in the 1,200 member AN language family marks a sharp departure from the approach adopted by Brandstetter, and followed by Blust (1988b). But it is not the only major departure from Brandstetter's method of operation. Perhaps even more importantly McCune (1985:1) notes that Brandstetter (and Alisjahbana, who simply cited him) "assumed a final -CVC root, while I divide the (normally disyllabic) root thus: CVC/ . . . /VC."

This difference of procedure leads to very different results in the recognition of recurrent sound-meaning correlations below the level of the morpheme. Drawing on the terminology of historical linguistics, but for a very different purpose, McCune (1985:71) describes collections of morphemes that in his view share a submorphemic sound-meaning association as 'subgroups'. To illustrate,

---

[1] Blust (1988a) was substituted for a different paper presented as a keynote address at the Third Eastern Conference on Austronesian Linguistics (TECAL), which was organized through the leadership of Richard McGinn, and held at Ohio University in Athens, Ohio on May 6 and 7, 1983. McCune (1985) appeared during the time that this transition between spoken address and published address took place.

https://doi.org/10.1515/9783110781694-002

he argues (1985:73) that "The first subgroup is 'go in: enter, penetrate', and as support for it he cites 1. *acap* 'get into s.t. deeply (e.g. stuck in mud), under water, flooded, soak', 2. *debap* 'plop', 3. *selulup* 'to dive, plunge', 4. *selusup* 'to penetrate, infiltrate, slip away, disappear', and 5. *sisip* 'penetrate', insert (needle, splinter), slip in (book mark), darn (socks), repair (roof), infix." Despite his opening statement that the 'root' in his analysis is the final -VC of a morpheme, this example (and those in many other 'subgroups') shares only a final consonant, in this case the voiceless bilabial stop -*p*, a segment that McCune sees as defining no fewer than eight distinct 'subgroups' that he labels a) through h), with semantic definitions including 'go in: enter, penetrate', 'go in, pierce', 'take in: eat', 'take in: breathe', 'take in: absorb', 'take in: contain', 'take in: contain: hold, catch', and 'out: come out, give out, emit' (1985:73–74).

The first problem with McCune's results is the rather diffuse nature of the proposed semantic connections he proposes. Definitions for the five terms that McCune assigns to his first 'subgroup' appear in the *Kamus Besar Bahasa Indonesia* (Tim Penyusun Kamus 1989) as follows: 1. *acap* 'masuk dalam-dalam' ('enter deeply, as dagger in flesh, feet in mud'); 'tergenap (oleh air)' ('flooded by water'); 'terendam (dl. air)' ('immersed in water'), 2. *debap* 'tiruan bunyi barang jatuh dsb.' ('the sound of something falling, etc.'); 3. *selulup/ber-selulup* 'menyelam dl. air' ('sink under water'); 4. *selusup* 'masuk dgn. sembunyi-sembunyi' ('enter silently or secretly'); 5. *sisip* 'menyelip (diantara dua benda atau di sela-sela sesuatu' ('to insert between two things or in the gaps between things'). While one might argue that there is a general sense of entering or penetration about each of these words (with the possible exception of *debap*), it is a sense that varies widely in its particular expression from one morpheme to the next. This contrasts rather markedly with the more specific semantics of -CVC roots such as *-Cuk 'to knock, pound, beat', *dem 'dark; overcast', or *-kaŋ 'spread apart, as the legs', which have exactly these senses or slight deviations from them in scores of languages and numerous non-cognate morphemes.

The second problem that some readers may have with McCune's results is that the -*p* that defines the above eight 'subgroups' has two historical sources. Because Malay, the source language for the artificial creation that is Bahasa Indonesia, underwent final devoicing, voiceless stop codas have both voiced and voiceless origins, and where etymologies are known, it can be shown in some cases that not all forms in the same subgroup reflect the same original final consonant, as in subgroup d), where Indonesian *isap* 'to smoke; to suck; to inhale' reflects PMP *qisep 'sucking, soaking up, absorbing', while *k-uap* 'yawn' reflects PAN *Suab 'yawn' (Blust and Trussel 2020). Although Alisjahbana (1949–1950) followed Brandstetter in recognizing a -CVC root, his results, like McCune's, suffered from the problem of conflating voiced and voiceless stop codas which have merged in Malay/Indonesian.

The third problem that some readers may have with McCune's results has already been anticipated in the preceding comment about the ambiguous voicing status of word-final stops in Indonesian, namely that he has restricted himself to a single language, and preliminary investigation suggests that trying to generalize his claims beyond Bahasa Indonesia does not produce convincing results. This stands in stark contrast to the -CVC root of Brandstetter (1910) and Blust (1988b), which shows close agreement across hundreds of languages, strongly confirming that it is a product of actual language history, and not of the imagination of a single investigator. In this connection it is worth noting that McCune (1985:18) appears to believe that synchronic and diachronic approaches to the study of submorphemes must lead to different results:

> We will examine the evidence for making the primary submorphemic cut *after* the medial consonant of the root (as this study does), as opposed to *before* (as Brandstetter does). As we do this, two facts must be clear at the outset. First, that we are comparing unlike things: Brandstetter's historical claims about the nature of roots in Proto-Austronesian with my claims about "lateral coherences" or synchronic word-families in modern Indonesian. Second, even viewed as a claim with synchronic consequences, Brandstetter's parsing and mine need not be seen as mutually exclusive. Later I will discuss various ways of "harmonizing" different commonalities highlighted by the different parsing systems.

In a footnote embedded within this quotation, McCune (1985:32–33, fn. 82) adds: "To use Matisoff's (1978:17, 18) terms, Brandstetter is looking at diachronic, crosslinguistic allofamy, while I am looking at synchronic, monolingual allofamy. In Matisoff's discussion, all four possible combinations of these terms are suggested, but in actual practice there is a tendency for an investigation of crosslinguistic allofamy to have diachronic claims as its goal."

Readers who are familiar with Brandstetter's work are likely to find this characterization surprising. First, Brandstetter never spoke of 'Proto-Austronesian', but only of 'Common Indonesian' and 'Original Indonesian' (Brandstetter 1916: xi). Second, he spoke both of "seeking the root in an individual language" and of "seeking the root by means of the comparison of languages." It is therefore inaccurate to claim that Brandstetter's investigation of the root in AN languages was based exclusively, or even primarily on a diachronic approach. Rather, he noted that submorphemic -CVC elements are found in individual languages, but that the investigation of roots would be incomplete if it were to stop there, given the extent of **agreement** across languages in the roots that they contain. It would be more accurate, then, to say that Brandstetter's approach was to let the data speak for itself, rather than forcing interpretations upon it, and that he saw no sharp line separating synchrony from diachrony in establishing the reality of -CVC roots. Rather than being a qualitatively distinct area of inquiry, to Brandstetter the diachronic study of the root in AN languages was an *extension* of the synchronic study, not something different from it. For any given language it is uncommon for more than two or three morphemes to contain the same -CVC root, but by widening the inquiry to include a larger number of languages the number of agreements (and hence the security of the inference) grows progressively stronger.

To sum up, McCune's experiment with Bahasa Indonesia is worth examining on its own, but it hardly presents a convincing case for the existence of -VC (or, more accurately, -C) submorphemic elements that have a consistent, generally recognized semantic value. As shown in Blust (1988b:55ff), although submorphemic sound-meaning correlations other than -CVC roots do exist in AN languages, the great bulk of convincing submorphemes are of the type first recognized by Brandstetter.

Nothofer (1990, 1991). Blust (1988b) was reviewed by several scholars, including Geraghty (1990), Kaufman (1990), Naylor (1990), and Nothofer (1990). Kaufman's review was by a non-specialist without insider knowledge of the languages, and was both brief and superficial. The other three were written by Austronesian specialists who were in a better position to appreciate the issues involved. Those by Geraghty and Naylor were overwhelmingly positive, while the review by Nothofer adopted a tone that made it difficult for the uninformed reader to know what contribution the book under review made to the growth of knowledge in the Austronesian field. The first sentence of Nothofer (1990) reads: "It is a pity that the quality of an interesting and detailed study such as Blust's book on what he calls "Austronesian root theory" suffers significantly because of an insufficient consideration of earlier works on this topic." The second paragraph then begins with "If Blust had studied works by authors such as W.E. Maxwell (1882), C.N. Maxwell (1936), and Wilkinson (1936), and if he had tried to test their hypotheses, he might have been able to realize that other recurrent elements than those posited by Brandstetter can be established." This is strong language, and it is important to state as clearly as possible: 1. what contribution Blust (1988b) made to scholarship on the Austronesian languages, and 2. how important the work of the Maxwells and Wilkinson actually is to the topic at issue.

With regard to the first point, the reader might well conclude from Nothofer's language that Blust (1988b) suffers from shortcomings that make its claims of little value. For those who have not read the book, it is essential to point out that it contains two major parts, the first on Austronesian root theory and the second (Sect. 6) on 'related matters'. The latter is further subdivided into 6.1. initial consonant symbolism, 6.2. medial consonant symbolism, 6.3. final consonant symbolism, and 6.4. Gestalt symbolism. Nothofer provides a perfunctory recognition of this division, but does not distinguish the parts, or comment on why 'related matters' is not included under 'root theory'.

Closer attention to the book he reviewed would have made it clear that 'root theory' is confined to an examination of the recurrent -CVC elements first proposed by Brandstetter in 1910, and 'related matters' deals with other types of sound symbolism, as is explicitly stated at the beginning of Sect. 6: "In addition to possessing roots, many AN languages exhibit a recurrent association of a single segment or a configuration of non-contiguous segments with a well-defined meaning. McCune (1985) discusses some of these for Malay/Bahasa Indonesia. Like roots, such meaning-related submorphemic elements ("submorphemes") pose a serious challenge to the view that the morpheme is the minimal meaningful unit in language."

Blust (1988b) sought to advance the study of the AN 'root' beyond the pioneering effort of Brandstetter in the following ways:

- First, Brandstetter (1916) recognized roots even if they are attested in a single language, as Old Javanese, or Karo Batak, while Blust (1988b) requires stronger evidence to control the play of chance. Discounting these examples, I count 40 roots that Brandstetter established with crosslinguistic support, as against 231 in Blust (1988b).[2]
- Second, Brandstetter (1916) recognized roots based on their proposed occurrence in just two non-cognate morphemes (e.g. p. 18, "Karo [Batak] *bebas* 'accustomed', 'Bisayan' *basbas* 'to accustom': root *bas*,") while Blust (1988b) requires at least four etymologically independent witnesses.
- Third, Brandstetter sometimes accepts cognate forms as etymologically independent, as in Karo Batak *ərdan*, 'Bisayan' *hagdan*, root: *-dan 'stairs', when both reflect PMP *haRezan 'notched log ladder', Karo Batak *api* 'fire', 'Bisayan' *apuy* 'erysipelas', root: *-pi/puy 'fire', when both reflext PAN

---

[2] The examples I find in Brandstetter (1916) are: p. 7: 1. *-lih 'choose'; p. 18: 2. *-kil 'to gnaw', 3. *-gal 'unchaste', 4. *-tat 'slow, slothful', 5. *-dan 'stairs', 6. *-nam 'joyful', 7. *-bas 'accustomed', 8. *-kut/kot 'to kindle a fire', 9. *-pi/puy 'fire, erysipelas', 10. *-buk.bok 'smoky', 11. *- məs/mos 'dissolve, wet', 12. *-lar/lag 'to shine', 13. *-sur 'satiated', 14. *-wət/wod 'sea'; p. 19: 15. *-pis 'thin', 16. *-saŋ 'gills', 17. *-ŋis 'angry', 18, *-laŋ 'alternate; interval', 19. *-səl 'repentance'; p. 20: 20. *-liŋ 'to roll', 21. *-pit 'to pinch', 22. *-dəm 'to close the eyes', 23. *-kar 'to open' (plus *-pis 'thin', and *-səl 'repentance', which were previously mentioned in connection with a different set of languages); p. 21: 24. *-bih 'surplus', 25. *-ruŋ 'enclosure', 26. *-liŋ 'to pour', 27. *-pat 'stretched out', 28. *-jak 'to stamp', 29. *-pas 'loose, free', p. 27: 30. *-pi 'dream', 31. *-li 'to buy', 32. *-ju 'to hit, to aim', 33. *-si 'contents'; p. 36: 34. *-liŋ 'to turn, spin', 35. *-liŋ 'to look', 36. *-liŋ 'word, sound'; p. 53: 37. *-suk 'to enter'; p. 53–54: 38. *-təm 'black', 39. *-lun 'wave', 40. *-lit 'skin' (plus the previously considered *-ruŋ 'enclosure' and *-pas 'loose, free').

*Sapuy 'fire', or Karo Batak *lawət*, 'Bisayan' *lawod*, root: *-wət/wod 'sea', when both reflect PAN *lahud 'downstream, toward the sea' (Blust and Trussel 2020). In Blust (1988b) cognate forms are carefully distinguished from EIMs.
- Fourth, Brandstetter interprets Old Javanese *luh* 'tears' as a monosyllabic root which "runs through nearly all the IN languages with the meanings 'to flow, to weep, tear'" (1916:21, Sect. 31). However, it actually reflects PAN *luSeq 'tears' with syllable reduction (Blust and Trussel 2020), and therefore has no etymological relationship to other words that end with a similar -CVC sequence, and is, in effect, an 'invented' root.
- Fifth, Blust (1988b) raised the fundamental theoretical question how roots (or phonesthemes in general) are transmitted across the generations. While the answer to this is straightforward in cognate morphemes, which are learned as part of the general vocabulary of a language, it is far from straightforward in non-cognate morphemes, where it is not the morpheme that is transmitted, but rather a more abstract sound-meaning association that cannot be extracted by contrast, but only by recurrence. Brandstetter never raised this question, or even seemed to be aware that it is a serious issue.

In short, despite the care that Brandstetter took in anticipating possible objections to his claims, it is easy to forget how much progress was made in Austronesian comparative linguistics between 1910 and 1988, not only with regard to new or improved descriptions of many languages, but also with regard to reconstruction and method. Apart from some brief remarks about controls on inference in a book that he believes "suffers significantly because of an insufficient consideration of earlier works on this topic," Nothofer seems oblivious to these advances in his insistence that this study is defective because it has not included references to four peripheral publications that deal with sound symbolism in a single language, Malay.[3]

---

3 The fourth is Collins (1979), which proposes that the initial CV- in Kedah Malay [dəɲūt] 'of a heartbeat or pulse', next to [ɲūtɲūt] 'of repeated "sucking" pain as of an infected wound' or [kəcăp] 'of repeated chomping on food', next to [căp] 'of lips smacking' are prefixes representing respectively 'preliminary or auxiliary perceptions in addition to the main perception' and 'repetition of that main perception.' It is hard to understand why this example is cited as undermining the claims in Blust (1988b), since there is clear evidence of contrast here, suggesting that *də-* and *kə-* are morphemes, not phonesthemes (although, if they are restricted to these forms, [ɲūtɲūt] and [căp] could be back-formations).

With regard to the second point, as noted in passing by Nothofer, Blust (1988b) in fact discusses several types of sound symbolism in AN languages that had never previously been noted in print, and are clearly distinct from the -CVC 'root'. These include the very high incidence of *k*- or *g*- in morphemes that refer to rubbing, scratching, scraping and the like, of *ŋ*- in morphemes relating to the oral or nasal area (published separately as Blust 2003a), of -*y*- in morphemes relating to the idea of swinging or swaying, and of -*l* in words meaning 'dull, blunt'. In addition, the term 'Gestalt symbolism' was proposed for configurations of segments and syllables that are recurrently associated with a meaning, as with the very large number of non-cognate words meaning 'wrinkled' that exhibit the highly distinctive configuration of being 1. trisyllabic, 2. beginning with a velar stop, and 3. having a liquid as the second consonant, or the collection of words referring to various types of stench that reflect the PAN template *(C)aŋ(e)CV(C).

Apart from objecting to 'root' as a description of Brandstetter's -CVC recurrent partials (following his former teacher, Isidore Dyen, he prefers to call them 'radicals', since 'root' is used by many linguists for an unaffixed base morpheme), Nothofer's main objection to Blust (1988b) is that it does not discuss the evidence for submorphemic CV- or -VCV- sequences that were proposed by the two Maxwells and R.J. Wilkinson, a type of data that he himself has elaborated well beyond these earlier sources. The following is a relatively complete listing of what Nothofer (1990) considers significant oversights in Blust (1988b):

1. The word-initial sequence *bu-* is uncommonly frequent in words meaning 'rounded' or 'globular'. Citing only reconstructed forms or non-cognate isolates, Nothofer lists 14 words under the label 'round', 32 under the label 'blown up (belly), swollen belly, fat belly, big-bellied, pregnant', 17 under the label 'posterior, buttocks, rectum, stern', 8 under the label 'hill, elevated ground, mountain', and 14 under the label 'lump, bump, hump'.

    In addition to this CV- sequence, Nothofer notes "two more radicals which have the form *bu*."[4] The first of these, which is documented in 32 EIMs, means 'rotten, mould', and the second, which is documented in 14 EIMs, means 'white'.
2. The medial sequence -*ela*- occurs in many unrelated morphemes in the meaning 'space between, interval, interspace' (18 examples).
3. The medial sequence -*iri*- occurs in many unrelated morphemes in the meaning 'line, row, file, trail, drag' (20 examples).

---

[4] Here and elsewhere Nothofer lists *bu* in a contextual vacuum, although it is clear from his examples that it is invariably word-initial.

4. The medial sequence -*awa*- occurs in many unrelated morphemes in the meaning 'wide open space' (10 examples).
5. The medial sequences -*abu*- and -*ebu*- occur in many unrelated morphemes in the meaning 'dust', or ashes', documented in 19 examples for -*abu*- and 15 for *ebu*- (plus a brief mention of -*apu*- 'dust, ashes' in 4 cases). In addition, these are found in words with the related senses 'fog, mist, smoke' in another 20 examples.
6. The medial sequences -*ua*- or -*uCa*- occur in many unrelated morphemes in the meaning 'vomit', documented in 20 examples for the former variant and 10 for the latter. Nothofer regards these as onomatopoetic (imitating the sound of retching).
7. The medial sequence -*ia*- occurs in many unrelated morphemes in the meaning 'cry, whine', documented in 8 examples that are also considered onomatopoetic.
8. The medial sequences -*ai*- or -*awi*- occur in many unrelated morphemes in the meaning 'hook; grasp with a hook', documented in 9 examples for the former variant, and 12 for the latter.

Nothofer (1991) continues his documentation of recurrent submorphemic sound-meaning associations in AN languages by listing five categories that do not conform to the -CVC pattern first observed by Brandstetter, namely: 1. Medial radicals of the shape V(N)(C)V, 2. Radicals consisting of three consonants, 3. Radicals occurring in the initial and in the final syllable of word bases, 4. Word bases formed by combining two radicals, and 5. Radicals in initial position (subsuming CVC- radicals and CV- radicals).

(1) The first category, medial radicals of the shape V(N)(C)V, is represented by:
    1) *-iCa- 'shine, sparkle, light, dazzle' (32 examples)
    2) *u(N)(C)u- 'protruding', subsuming 'headland, point', 'mouth, labial orifice', 'lips', 'nose', 'beak, snout' and 'horn' (57 non-duplicated examples)
    3) *aCa 'mouth' (14 examples), and *aŋa 'agape, wide open', with 7 examples, one of which duplicates a citation under 'mouth'
    4) *u(N)(C)u- 'blunt, dull' (37 examples), plus 'dullness of wits' (27 examples)
    5) *u(C)u 'to mumble' (29 examples, considered to be onomatopoetic)
    6) *u(C)a 'come out, open up, take off', intended as a somewhat questionable cover term for 'bubble up' (15 examples), 'foam' (11 examples), 'spit, spittle' (16 examples), 'open, loosen, untie' (34 examples), 'yawn, gasp' (12 examples), 'wide open, wide, hole, hollow' (19 examples), come out,

outside' (18 examples), 'peel' (13 examples), and 'lever, dibbling stick' (12 examples)
7) *a(N)(C)u 'mix, knead' (27 examples)
8) *i(N)C)i 'tooth, show teeth, jagged' (20 examples)
9) *e(N)(C)e 'drown' (11 examples, treated separately on p. 253)

(2) The second category, radicals consisting of three consonants, is represented by:
   1) the formula *k .. (N)C .. p 'blink, flash', with 20 examples (*kilap 'shine', *kesap 'blink', etc.)
   2) the formula *l .. (N)c/s .. (C) 'slip, slide, glide', with 13 examples (*lancar 'slipping along smoothly', *lucut 'slip away', etc.)
   3) *l .. (m)p .. t 'fold', with 5 examples (*lipet 'fold', *li(m)pi(Ct) 'press between two flat surfaces', etc.).

(3) The third category, radicals occurring in (both) the initial and final syllable of word bases, is represented by:
   1) *kuŋ 'throat' (first syllable in 5 examples, last syllable in 4 others)
   2) *pu 'master, lord' (first syllable in 4 examples, last syllable in 3 others)
   3) *ba 'carry' (first syllable in 5 examples, last syllable in 4 others)
   4) *til 'small protruding part' (first syllable in 6 examples, last syllable in 6 others)
   5) *pap 'flattened' (first syllable in 4 examples, last syllable in 7 others)
   6) *diŋ 'hear, ear' (first syllable in 2 examples, last syllable in 4 others)
   7) *li 'twist, turn, wind, whirling' (first syllable in 32 examples, last syllable in 18 others)
   8) *teg 'hit, beat, chop up' (first syllable in 3 examples, last syllable in 6 others)
   9) *teb 'prune, cut down, graze' (first syllable in 3 examples, last syllable in 7 others)

It should be noted that several of these (*-til, *-pap, *-teg, *-teb) are roots proposed in Blust (1988b) that Nothofer assigns to initial syllables as well as final syllables. As such, they are not additions to the inventory of new 'radicals', but are used as evidence against the claim in Brandstetter (1910) and Blust (1988b) that monosyllabic submorphemic roots are invariably, or almost invariably word-final.

4) The fourth category, word bases formed by combining two radicals, is based entirely on roots that appear in Blust (1988b), and which in Nothofer's view, can be seen as combining in forms such as *lu(ŋ)kuq 'bend, curve', which he

analyzes as *-luŋ 'bend, curve' + *-ku(q) 'bend, curve'. In view of the facts that 1. *-ku(q) in Blust (1988b) is found in 28 EIMs, and none of the others can be analyzed in this way, and 2. some languages reflect *lukuq rather than *luŋkuq, this claim appears fairly arbitrary. In any case, it adds no further radicals to the stock proposed by Nothofer.

5) The fifth category, radicals in initial position, is divided into two subsets: CVC- radicals, and CV- radicals. The first is represented by two cases and the second by six:
   1) *teg 'firm, bolt; sturdy' (9 examples)
   2) *lem 'soft, weak, moist, tired' (18 examples)
   3) *lu 'soft, soft matter' (33 examples meaning 'mud' or 'muddy', 5 examples meaning 'wallow in mud', 25 examples meaning 'soft, tender, ripe', 9 examples meaning 'grease, oil, besmear')
   4) *le 'moist, mud' (13 examples)
   5) *le 'tired, weary, slack, give (of a bridge)' (17 examples)
   6) *tu 'heel, stump, base, prop, support' (24 cases)
   7) *su 'comb' (15 cases)
   8) *si 'side, edge' (70 cases)

Note that some of the supporting examples cited by Nothofer are not etymologically independent, as Banjarese *lubur* 'mud', Cebuano *lubug* 'murky, muddy', both of which could reflect PWMP *lubuR, or Buginese *lura*, Minangkabau *luda?* 'mud', both of which could reflect PWMP *ludaq.

Nothofer's contribution to the discussion of submorphemes in AN languages is valuable and important, especially his demonstration that recurrent -V(N)(C)V- sound/meaning sequences may crosscut -CVC roots, and so allow the recognition of different patterns of sound-meaning association across the same phonological segments, as in *-iCa- 'shine, sparkle, light, dazzle', next to *-laC, *-lak, *-laŋ, *-lap, and *-law when following *i, all referring to the same semantic field. However, it cannot be emphasized too strongly that such submorphemic sequences do not provide evidence *against* the -CVC root of Brandstetter. As he himself notes (1991:250–51), in a much more conciliatory tone than that with which he began his 1990 review, "Blust (1988[b]) clearly shows that radicals have to be recognized as occurring in the final syllable of word bases. Particularly convincing are those for which the supporting evidence is overwhelming, for example, pit 'press, squeeze together; narrow' or keC 'adhesive, sticky'."

Virtually every case that Nothofer cites fits into the category of 'related matters' in Blust (1988b), which dealt with initial consonant symbolism, such as the high frequency of *k*- and *g*- in words that refer to rubbing, scratching, scraping, etc., or the high frequency of *ŋ*- in words relating to the oral/nasal area, with medial consonant symbolism, as the high frequency of -*y*- in words that relate to swaying, swinging, etc., with final consonant symbolism, as the high frequency of -*l* in words that mean 'blunt, dull', or the striking patterns of Gestalt symbolism in words meaning 'wrinkled' and the like, which are often trisyllables with an initial velar stop and a liquid as the second consonant, or words referring to noxious odors which commonly fit the template *(C)aŋ(e)CV(C). What Nothofer's proposals do, then, is to expand the types of examples that Blust (1988b) treated under 'related matters', rather than demonstrate that the -CVC root is invalid.

Moreover, despite their value in showing that alternative analyses of submorphemic sound-meaning associations are possible in the vocabularies of AN languages, there are four major problems with Nothofer's proposals to diminish the role of the -CVC root first proposed by Brandstetter:
- Variable vocalism in the penult of -V(N)CV- patterns
- The root as onomatope
- The root in reduplicated monosyllables
- The far greater frequency of -CVC roots

The first problem encountered in assessing the alternative -V(N)CV- patterns proposed above, arises from the appearance of variable vocalism in the penult of the relevant morphemes. As noted above, Nothofer (1991) acknowledged the existence of -CVC roots where "the supporting evidence is overwhelming, for example, pit 'press, squeeze together; narrow' or keC 'adhesive, sticky'." He does not state his reason for considering the evidence for -CVC in these cases as "overwhelming", but one might suppose that it is because they are the two most richly exemplified of the 231 roots cited in Blust (1988b), with 48 etymologically independent attestations for *-pit, and 44 for *-keC. While robust attestation is a reasonable basis for assuming why Nothofer accepts these roots as secure evidence for -CVC, it does not appear to be critical. More important is that a -VCV- alternative to -CVC will work only where the penultimate vowel is constant, and this becomes progressively less likely as the number of EIMs increases. This is true of *-pit and *-keC, which permit all four PAN vowels in the penult, and hence do not allow a phonetically constant -VCV- alternative, but it is also true of nearly all roots that appear in at least 8 EIMs, which includes 147 of the 231 cases in Blust (1988b). In effect, Nothofer is able to support his argument for radicals such as *-iCa- 'shine, sparkle, light, dazzle', by appealing to the statistical dominance, rather than to the exclusive occurrence of a favored penultimate vowel.

Considering only the forms that I assign to *-laC, *-lak, *-laŋ, *-lap, and *-law, the penultimate vowel has the following values:

*-laC: i = 6, e = 2, hence /i/ = 75%
*-lak: i = 12, consonant = 3, a = 1, u = 1, hence /i/ = 70%
*-laŋ: i = 3, consonant = 1, u = 1, e = 1, hence /i/ = 50% (but is still dominant)
*-lap: i = 12, consonant = 9, e = 8, a = 2, u = 2, hence /i/ = 36% (but is still dominant)
*-law: i = 9, e = 3, hence /i/ = 75%

Extracting a -VCV- pattern from this data is possible, but only by selectively choosing favorable cases (42 with these five roots), and eliminating unfavorable ones (34 with the same set). Viewed in this way Nothofer's record of collecting large numbers of morphemes with a medial sequence -iCa- that mean 'shine, sparkle, light, dazzle' is less impressive than it might appear at first sight.[5] This is not to deny that in this particular case /i/ is the dominant vowel in the penult, while in morphemes selected without respect to meaning /a/ is more frequent in this position. This observation is still in need of explanation, but the case for a submorpheme *-ila-* in the above subset of -iCa- cases meaning 'shine, sparkle, light, dazzle', is clearly weakened by a recognition of variable vocalism in the penult of -VCV- patterns.

The second problem with Nothofer's attempts to marginalize the -CVC root arises from onomatopoetic roots which may occur in isolation (as pointed out in Blust 1988b, this is virtually unknown with non-onomatopoetic roots). While these are not numerous, they permit only one interpretation, namely that the same phonetic sequence with the same meaning in a CVCVC morpheme is a -CVC root. Thus, given PMP *bak 'clap! smack!', PMP *be(m)bak 'to slap, beat on', PAN *Cebak 'to smack against', PWMP *debak 'to pound, thud', PWMP *lebak 'to pound, thud', Dairi-Pakpak Batak gəbak-gəbak 'sound of wings flapping', Javanese kə-kəbak 'beating of wings', it would be possible, but pointless to posit -e(m)ba- as a 'radical', not only because other instances of this root appear in morphemes in which the penultimate vowel is /a/, /i/ or /u/, but most crucially because PMP *bak (based on Tboli bak 'bark of a dog', Ngaju Dayak bak 'splash!',

---

[5] Nothofer (1991) acknowledges this as a problem with his 1990 proposal that *-bit 'hook, clasp; grasp with fingers' and *-wit 'hook-shaped' in Blust (1988b) can be better modeled by recognizing a(N)(C)i 'hook', since many examples that contain -bit or -wit have penultimate vowels other than /a/. He counters this admission by noting that I cannot account for the many -a(C)iC words for 'hook' in AN languages in which the penultimate consonant is not *b or *w, or the final consonant *t, as with Atayal nayip, Banjarese bait, or Ilokano lawin 'fishhook'.

Malay *bak* 'sound of a smack', Karo Batak *bak* 'sound of clapping, of horses' hooves, etc.', Dairi-Pakpak Batak *bak* 'sound of a large object falling on a wooden floor', Ngadha *ba* 'to smack, clap wth the hand', etc.) gives us the root in isolation.

This leads us inevitably to the third and most damaging problem for Nothofer's attempt to marginalize the -CVC root, namely its occurrence in reduplicated monosyllables. As noted already, most -CVC roots are not onomatopoetic, and therefore may not occur in isolation. However, both onomatopoetic and non-onomatopoetic roots occur in reduplications, and – unlike onomatopes – these are quite numerous. Brandstetter (1916:43–47) noted that many of the -CVC sequences that he characterized as *Wurzeln* occur not only with a CV-formative that has no recurrent meaning, but also as reduplications, as in *kapkap* 'kite (predatory bird)', next to *taŋkap* 'to seize'. Blust (1988b) noted the same tendency – that many (not all) roots that can be securely established in CVCVC morphemes also occur as reduplicated monosyllables, and these can only be interpreted as the doubling of a -CVC root – no other analysis is possible.

The present corpus of 406 -CVC roots includes at least 186 unambiguous cases of CVCCVC reduplications attested in either reconstructed forms or morpheme isolates. To avoid possible controversy, I have treated only $C_1V_1C_2C_1V_1C_2$ bases as reduplications, ignoring $C_1V_1C_1V_1C_2$ bases as potentially containing an unrelated CV-, even in languages that regularly reduced medial clusters. Even with these strict procedures, over 46% of all 406 roots proposed here appear in one or more languages with full reduplication of a -CVC submorpheme, and *none* of these can have an alternative -VCV- analysis. This means that although alternative analyses may be possible for some submorphemic sound-meaning correlations, nearly half of those cited here can *only* be analyzed as containing a -CVC root.

Finally, we come to frequency. As seen above, Nothofer (1991) proposed the following radicals: (1) medial radicals of the shape V(N)(C)V (9 cases), (2) radicals consisting of three consonants (3 cases), (3) radicals occurring in (both) the initial and final syllable of word bases, (9–4 = 5 cases), (4) his fourth category, word bases formed by combining two radicals adds nothing new, since both radicals were already recognized, and (5) radicals in initial position (8 cases). In all, then, he has argued for the recognition of 25 submorphemic sound-meaning associations (35 if we include the 10 examples presented in Nothofer 1990, minus duplications). This compares with 406 -CVC roots in the present collection, or over eleven times the number he claims. If all, or many of these 406 roots are products of accident or faulty analysis one must explain why they are so much more numerous than any other type of submorphemic sound-meaning association in AN languages. It is precisely this property that brought them to the early attention of Brandstetter, who had no particular reason to recognize the more marginal kinds of submorphemic sound-meaning correlations that Nothofer treats as central.

Zorc (1990). This paper, published in an edited volume dedicated to demonstrating the universal applicability of the comparative method, consists largely of a step-by-step dissection of the procedures followed in Blust (1988b). However, it departs in certain ways from the methods used in that publication. With regard to terminology, Zorc notes (as does Nothofer 1990:133), that objections might be raised against the use of 'root' for submorphemic elements, since 'root' is more commonly used in English to refer to an unaffixed base morpheme. However, unlike Nothofer, who insists on rendering Brandstetter's *Wurzeln* as 'radical', he tries to accommodate everyone in the title of this paper, noting (1990:178) that "if used consistently in a phrase, such as 'monosyllabic root', 'root candidate' (etc.), the meaning should be clear, and ambiguity avoided."

More substantially, Zorc (1990:179) notes that: "Dyen and Zorc (personal conversation in 1974) outlined the following criteria for establishing a monosyllabic root:
1. Find a doubled monosyllable and a form with an established affix, e.g. *puk + puk 'beat, pound', *ka-púk 'beaten fibers'.
2. Look for ineluctible parallelism of meaning, e.g. *suk + suk 'put in or on', *pa-suk, *ma-suk 'enter'.
3. Identify partial reduplications: *CV + CVC – *su + suk 'prick, pierce'.
4. Identify any instances or resyllabification, i.e. CVC + VC > *CV.CVC, especially if *CVC + -VC (suffix). Although this has not been found to be as productive as final CVC types, note the following possible interpretations: *taŋan 'hand' = *taŋ 'grasp' + *-an, *paRaw 'hoarse' = *paR 'id.' + *-aw, *bakíʔ 'frog' = *bak 'pounding noise' + *-iʔ.
5. Identify any binding of bases (i.e., compounds), e.g. rak + suk 'put on', *ruk + sak 'destroy'.
6. Establish splitting of bases with an epenthetic laryngeal (i.e. *CVC > *CVʔVC), or *CVhVC, e.g. *suʔuk 'enter' (*suk), *piʔət 'narrow' (*pit) or a semivowel, e.g. PHN *yak > iyak 'cry' or PPH *siw 'chick' > siyuʔ (cf.: PPH *siw + siw)."

An important point not touched on seriously by other writers is whether all roots were originally *-CVC, since the reflexes of some of these in various Philippine languages show a root-internal glottal stop that implies an earlier disyllable. To his credit Zorc leaves open the question whether the pre- or postconsonantal glottal stops that are associated with monosyllabic roots in a number of Philippine languages are innovations or retentions. If they are retentions, the glottal stop cannot reflect PAN *q where Kalamianic and Bilic witnesses fail to support this phoneme. On the other hand, if they are innovations either in Proto-Philippines, or in the histories of individual Philippine languages, they represent what would appear to be a sporadic process of laryngeal epenthesis, in violation of the Regularity Hypothesis. It is worth pointing out that much the same problem appears in trying to explain phonemic stress in Philippine languages, since in this matter, too, there is no external support for assigning this feature to a higher-level proto-language. Until this matter can be definitively resolved, I assume that the glottal stop in forms like Hiligaynon *yábʔuk* 'dust, dirt' (root: *-buk 'to decay, crumble; powder'), Kankanaey *gabʔún* 'to fill up (with earth, etc.), to cover' (root *-bun 'to heap up, cover with earth; collect, gather'), or Cebuano *sakáʔaŋ* 'to totter under a heavy weight with the legs spread far apart for balance' (root: *-kaŋ 'to spread apart, as the legs') is historically secondary. I assign it to PPH reconstructions where a *q is ruled out by diagnostic witnesses, but I also reserve the option of concluding that its appearance in many forms postdated the breakup of Proto-Philippines. For higher-level reconstructions I posit only monosyllabic roots except where an apparent root appears as a disyllabic free form, as with *-pit-$_2$ (or *piqit?) 'to press, squeeze together; narrow', which must include the longer variant to account for the disyllabicity of e.g. Casiguran Dumagat *piʔít* 'tight, narrow, crowded', Tagalog *piʔít* 'squeezed, pressed tightly between two persons or things', Malay (Brunei) *pihit* 'to press down; to weigh down', Kadayan *pihit* 'to pinch', Banggai *piit* 'shut, of the eyes'. To be consistent with this decision, I have revised the shapes of higher-level reconstructions (i.e. those ancestral to PPH) by eliminating root variants with medial *q that are unattested as disyllabic free forms (hence PWMP *Rab(e)qun, which included a disyllabic root only to accommodate Kankanaey *gabʔún*, is now written *Rabun, PWMP *cep(e)qak 'to crack, split, break', which included a disyllabic root only to accommodate Cebuano *súpʔak* 'to split s.t. lengthwise, esp. along the grain; to break off, said of tree branches' is now written *cepak, etc.).

Another useful part of Zorc's review of the submorpheme problem in Austronesian is his discussion of the CV- of CV(N)CVC bases in which the last syllable is a root. Like Brandstetter, who called them 'formatives', Blust (1988b) treated these CV- sequences as a meaningless residue left over after the extraction of the root, much like the *-eam*, *-isten*, or *-itter* of the English words *gleam*, *glisten* or *glitter*. However, there is always the temptation to see the formative as a morpheme, and in some cases it corresponds in form (though rarely in function) to an established prefix. Again, Zorc wisely leaves this matter open rather than trying to force a meaning upon these CV- sequences.

Additional useful features of this paper are: 1. A brief discussion of how the recognition of monosyllabic roots can disambiguate the shapes of some reconstructed morphemes, as where *-cik 'to fly out, splash, spatter' can be confidently established based on attestation in test languages for the *c/s distinction, such as Iban, Malay, or Javanese, and then extended to reconstructions like PMP *becik 'to spatter, fly off in all directions', where none of the languages that support this form distinguishes *c from *s, and 2. A passing remark on how the reconstruction of disputed proto-phonemes such as *c, *g or *r (Wolff 1974, 1982) can be strengthened when these segments are found in morphemes that need not be cognate.

Somewhat less helpful is Sect. 7, titled 'Zorc's problem – the laryngeals', with no explanation for the uninformed reader as to what 'Zorc's problem' is, other than that it has something to do with the proposed need to reconstruct word-final *-ʔ in addition to *-q, *-h and final vowels in monosyllabic roots. Zorc follows his discussion with two appendices, the first a list of abbreviations, and the second a comprehensive list of 256 proposed monosyllabic roots that are said to come from three sources: 1. Blust (1976), 2. Blust (1988b – 'in press' at that time), and 3. Zorc's own files. The coding used shows the following credits, where 'B' = Blust only, Z = Zorc only, and BZ = Blust and Zorc:

B	BZ	Z
69	82	188

This result is numerically odd, since the Blust citations, whether alone or in combination with Zorc, amount to 151 cases, yet 231 roots are proposed in Blust (1988b), implying that 80 of these were either rejected without explanation, or for some reason overlooked. More seriously, no supporting evidence is given for any of Zorc's 188 proposed roots, a point also made by Nothofer (1991:223). Many of these may prove to be valid, as a number of them had already turned up in the ACD when I began to contemplate writing this manuscript. However, the absence of supporting evidence is a serious shortcoming, as without examples the reader has no way of judging how many supporting examples were used to propose any given root, some of which may suffer from the same shortcoming as Brandstetter roots that were based on just two EIMs.

Potet (1995). Like the British writers in colonial Malaysia, and McCune (1985), but unlike Brandstetter, the French linguist Jean-Paul G. Potet has directed his attention to the identification of monosyllabic roots in a single language, in this case Tagalog. As seems to be common in cases where a single attested language is investigated, the absence of external controls against the play of chance appears to give greater freedom to the play of the imagination.

Potet begins his treatment of Tagalog roots by saying (1995:345) "I discovered Tagalog monosyllabic roots in 1979 while trying to find mnemonic devices to remember Tagalog words with the correct stress. I soon realized that the phenomenon deserved special study, and included the results in my research on Filipino morphology." He follows this up by stating his belief that monosyllabic roots can be found in individual attested languages without the use of comparative evidence, implying, like McCune, that this approach departs from that of earlier researchers, even though Brandstetter (1916) contains separate sections titled 'Seeking the root in an individual language', and 'Seeking the root by means of the comparison of languages'.

In commenting on terminology Potet then remarks "I have consistently used the word "root" (Fr *racine*, Ger *Wurzel*) as in Indo-European studies. A root is generally defined as a monosyllabic morpheme obtained after all additional items have been eliminated (Marouzeau 1951:194). This definition has been adopted by most Austronesianists, with the exception of Dyen and Nothofer, who propound "radical" instead. A radical (Fr *radical*, Germ *Stamm*) being built on a root, I think it advisable not to use the latter for the former and vice versa." Following this he launches into a highly idiosyncratic theoretical discussion of Tagalog phonology that need not concern us further.

To give some idea of the types of submorphemic sound-meaning correlations that Potet proposes for Tagalog we need only consider a few examples. The first that he gives is **tay** 'death', said to be present in *patáy* 'dead', *bitáy* 'hanging', *himatáy* 'swoon' and *dátay* 'bedridden'. While these four terms exhibit a sufficient degree of semantic similarity to pass as a root in Tagalog, there is little support for it in other languages, which reflect *-Cay only as the final syllable of PAN *aCay 'death', and a variety of morphological derivatives (Blust and Trussel 2020). Potet's second example, which he uses to make the point that a root may be inferred in Tagalog once it is found in at least three words, is **lat** 'tree bark > record', seen in *aklát* 'book', *súlat* 'writing', and *úlat* 'report'. Already, with this example, the reader may begin to sense that the material is being manipulated rather than showing genuine submorphemic sound-meaning correlations, since paper has only been known by Tagalog speakers since the arrival of the Spanish. Potet's third example is sure to strengthen this impression of arbitrariness in his treatment of semantics, as he bases the root **bay** 'shore' on *baybáy* 'shore', *balaybáy* 'going along a shore or a river bank', *hibaybáy* 'seaside or lakeside area', which then morphs into 'railing', with *gabáy* 'bannister', *balabáy* 'open gallery with a balustrade on the side of a house', then to 'whoever or whatever keeps you on the right track', with *gabáy* 'guide', and *patnúbay* 'director', then to 'companion', with *sabáy* 'together', *alakbáy/akbáy* 'with one's arm over one's companion's shoulder', and *ábay* 'best man at a wedding', and finally to 'along something', with *agabáy/agapáy* 'parallel'.

Many of Potet's other examples of Tagalog 'roots' are equally unconvincing on semantic grounds. It is worth noting that within this semantically quite variable cluster of forms that are said to share a common root, only the last sense is accepted by Zorc (1990:188), who has *bay 'be together', and who by implication (since he has inspected words ending in -bay in Tagalog) rejects the other five, with balaybáy and hibaybáy probably being considered morphological derivatives of baybáy in the history of Tagalog. Likewise, as can easily be determined by inspection, none of the roots proposed in Blust (1988b) or in the present collection come even remotely close to making the same demands on the reader's willingness to suspend disbelief regarding semantic relationship as many of Potet's examples.

Wolff (1999). Unlike the other studies considered here, Wolff's is focused specifically on Proto-Austronesian (although many of his proposed 'roots' fail to meet that standard). Oddly, he appears to be unaware that many roots are found in morpheme isolates in the modern languages, and he consequently treats the topic exclusively as one of inheritance from a common ancestor, with no stated appreciation of the problem of transmission that inevitably arises with the recognition of roots in morphemes that do not belong to an established cognate set.

Early in his paper Wolff (1999:140–41) acknowledges his indebtedness to other scholars:

> This work builds on the work of predecessors: Brandstetter (1916), McCune (1985), Blust (1988[b]), Zorc (1990), and Nothofer (1990), but the basic data are from my own files and cited in the transcription of the phonology which I believe characterized PAn. All of these except McCune (1985) reconstruct monosyllabic roots. I do not discuss all of the forms which are adduced as monosyllabic roots in these studies, for I believe that the principles for recognizing monosyllabic roots presented there do not allow us to distinguish what can be established unequivocally from what is speculative.

While 43 of the 120 roots in Wolff (1999) were proposed and fully documented in Blust (1988b), these may well have been in his personal files before that work appeared. What is most disconcerting about the statement quoted above, however, is its claim that earlier studies failed to enunciate principles which allow an independent determination whether the roots that were proposed are products of principled research, or merely the result of speculation. Nowhere in his paper does Wolff discuss method, and he seems oblivious to the pains taken in Blust (1988b) to rule out chance as an alternative to history in determining the confidence level of a given root. With this theoretically rather feeble beginning, he then launches into an inventory of what he calls 'roots'.

The first thing that anyone familiar with the work of Brandstetter is likely to notice in Wolff's idiosyncratic treatment of the data is his failure to distinguish 'root' from 'morpheme'. His second example (1999:149–50) is *bay 'woman', which he cites first in unaffixed form, and then infixed with *-in-, and reduplicated and suffixed with *-an. However, all of these forms are reflexes of PAN *bahi 'female' (Blust 1982, Blust and Trussel 2020), with various affixes. Nowhere in his treatment does Wolff appear to be aware that the -CVC root in AN languages as treated by Brandstetter, Blust (1988b), and most other writers, is the kind of submorphemic sound-meaning association commonly called a 'phonestheme', which is clearly distinguished from the morpheme in the latter work, as it is in the present work.

As the reader moves on to other examples the discussion does not become any more enlightened. The next example that is not also in Blust (1988b)—in other words, the next example that seems to be completely original with Wolff—is given as *buñ 'fontanelle'. Setting aside Wolff's unique orthography (palatal nasals do not occur in non-derived final position in any attested AN language, and have never been reconstructed in this position by anyone else), he cites Cebuano *hubun, hubun-hubun* (which he speculates is a metathesized form of *buhun), Old Javanese *əmbun, əmbun-əmbun-an*, Malay *ubun-ubun*, Makassarese *ubuŋ*, and various reflexes of *bunbun 'fontanelle'. However, all of these together amount to just three etymologically independent morphemes (*hubun, embun, bunbun*), and so fall short of the standard set in Blust (1988b), in which root candidates were considered unlikely products of chance only if found in a minimum of four etymologically independent morphemes. Again, despite his blanket statement about the work of his "predecessors" not adequately distinguishing between valid identifications and speculations, Wolff himself proposes no objective criterion for evaluating the role of chance in identifying valid roots. Ironically, in some ways this basic shortcoming hardly seems to matter, since he shows little evidence of understanding the morpheme/phonestheme distinction, which is critical to any discussion of this topic.

To show that Wolff's failure to distinguish morphemes from submorphemes is not an error in a single form I will move on to the next two 'roots' which are original with himself: *Riq 'sword grass: *Imperata cylindrica*', and *kan 'eat'. The first of these is represented by three doublets in Blust and Trussel (2020): PMP *eRiq, PMP *keRiq, and PAN *Riaq. The only root that can be extracted from this set of forms (or from the similar material cited by Wolff) is *-Riq, and again this fails to meet the minimum requirement of attestation in four EIMs as established in Blust (1988b), and followed in the present work. The second 'root' is PAN *kaen, the base morpheme for the verb 'to eat', which occurs in a number of different affixed forms. Once more, Wolff treats the affixed forms as though they provide further support for a monosyllabic root *kan, when all that the Comparative Method reveals is a single disyllabic base *kaen, and various parts of the morphological paradigm which it hosted.

I will choose only one more example to illustrate the serious shortcomings of this work, as otherwise the discussion would become excessively repetitive. Next to *-lit 'to wind, twist', proposed with 8 etymologically independent attestations in Blust (1988b), and 12 in the present work, Wolff proposes *-lid in the same meaning. His sole reasons for proposing *-lid rather than *-lit are Bolaang Mongondow *lilid* 'to roll, as over the ground', which he leaves unglossed, with a 'note-to-self' which reads '(look up meaning)', and Old Javanese *pulir* 'turning round, whirling, spinning', next to forms meaning 'twist, wind around', etc. with unambiguous reflexes of *-t in Manggarai (*lilit* 'encircle'), Javanese (*lilit* 'coil around') and Sundanese (*lilit* 'wind, bind around'). Entirely apart from this contradictory collection of evidence Wolff states (1999:153) that "This root is reflected as a monosyllable in Oceanic languages," and as evidence he cites Fijian *vaka-lii-taka* 'to twist rope taut with stick', Samoan *lii* 'sennit fastening to attach outrigger boom to outrigger', where vowel doubling is used to mark length. While I do not wish to categorically deny that *lii* in Fijian and Samoan could be a reflex of *-lit, the latter is attested as a root only in CVCVC morphemes or CVCCVC reduplications, and the probability that this CV morpheme in Fijian and Samoan is the same submorphemic sound-meaning association as the *-lit documented in this book is bound to make the reader wonder what prompted Wolff's remark that, unlike the work of his "predecessors," his method of operation allows us "to distinguish what can be established unequivocally from what is speculative."

Kempler Cohen (1999). This is the only book-length treatment of submorphemic roots in AN languages since the appearance of Blust (1988b), and as such the reader is entitled to expect it to make a major advance in our understanding of the subject.

Among the major claims made in this book – important enough to be featured on the back cover as an advertisement for its contents – are: 1. That "all wordbases in Proto Austronesian and its early descendants were coined exclusively from CVC morphemes", and 2. That "All wordbases in the most common form, CVCVC, that have been analysed otherwise by other writers are here analysed as having been coined by merger of two CVC morphemes, i.e., by overlap of the final of the first and the initial of the other so that these two (nearly-) identical consonants are expressed as one."

The first of these claims goes well beyond Brandstetter or Blust (1988b) in maintaining that *all* base morphemes that are widely shared by AN languages contain roots. Such a claim entails that CVCVC disyllabic bases such as PAN *laŋiC 'sky', *panas 'warm', or *beRat 'heavy', among many others, not only contain a previously unrecognized -CVC root, but that the traditionally residual *la-, *pa- and *be-, etc. is also a root that has been reduced by consonant fusion (called 'merging' by Kempler Cohen 1999:6). This is a very strong assertion, and given the amount of solid comparative work that had been done prior to Kempler Cohen's book without recognizing anything like this, and his own limited qualifications as an Austronesianist, one is bound to approach it with skepticism.

Kempler Cohen begins his exposition by redefining the morpheme in linguistic theory, and the root as used by all other investigators. In his words (1995:1), the morpheme is: "A phoneme or array of phonemes that
(i) conveys a particular fixed meaning wherever employed as a (or the sole) component of a wordbase or affix,
(ii) needs no additional phoneme(s) to convey its meaning, and
(iii) (in the case of an array) fails to convey its meaning if any member phoneme is deleted, except where the resultant vacancy has effect as an allophone (phonemic zero) of the deleted phoneme."

By contrast, a root in his definition is "A morpheme used as the basis for the coining of another morpheme, by phonetic and semantic alteration."

On the next page he makes it absolutely clear what he means: "The fact is that both Blust's "root" and Nothofer's "radical" are simply what linguistic parlance has long designated as "morpheme." For the sake of consistency, this latter term will generally be substituted for theirs in discussions below, except of course in quotations from their works, and in some cases where the distinction is immaterial (as in the well-established phrase "root theory")."

Virtually any linguist reading the preceding passage will find it jarring, since it totally ignores the need for contrast in distinguishing the morpheme from the phonestheme. Reduced to its essentials, Kempler Cohen's novel definition thus labels such sequences as -pit in Malay (h)apit 'pressure between two disconnected surfaces', capit 'pincers', mən-cəpit 'to nip', dəmpit 'pressed together, in contact', (h)impit 'squeezing pressure', or -ket in Kankanaey busíkət 'clay; the thick, gluey black earth of rice fields', níkət 'resin', paŋkət 'sticking, adhering, cleaving' as morphemes, no different from any commonly recognized free morpheme or affix. Since neither of these bound elements (nor any other in Blust 1988b) can be defined by contrast, this alters the commonly accepted definition of 'morpheme', and places e.g. the gl- of English glare, gleam, glimmer, glisten, glitter, glow on the same footing as -s 'plural'.

Given this idiosyncratic (and frankly, dysfunctional) definition of the morpheme, one must wonder how Kempler Cohen defines 'root', since it figures in the title of his book. As he puts it, a 'root' is "A morpheme used as the basis for the coining of another morpheme, by phonetic and semantic alteration." Morphemes can thus be embedded within larger morphemes, and they are then called 'roots' (as building blocks of larger morphemes). While this appears to blur the root/morpheme distinction, this is only the beginning of what many will see as an impressively disciplined exercise in chaos, since we cannot forget that *every* component of the 'root' is said to be a 'morpheme' ("all wordbases in Proto-Austronesian and its early descendants were coined exclusively from CVC morphemes"), meaning that *(h)a-, ca-, cə-*, etc. in Malay are also meaning-bearing elements. More precisely, in AN languages, the semantically empty CV- 'formative' of Brandstetter is a *reduced* CVC- morpheme that has lost its final consonant by "merging" with the initial consonant of the last syllable.

There is no need to pursue what Kempler Cohen (1999:3) calls 'The underpinnings of the uniform-root thesis', which essentially elaborates on the points already made. In support of his case he cites examples such as pe**lay** 'to weaken', ku**lay** 'to wilt', **lay**as 'to weaken', **lay**a 'to wobble', and **lay**u 'to wither'.[6] The reader is left wondering how such examples provide evidence that every wordbase is built "exclusively from CVC morphemes," since all that is demonstrated here is that the phoneme sequence **lay** can appear either word-finally or word-initially in wordbases of similar meaning, whereas the wording of Kempler Cohen's claim asserts that the non-bolded parts of these (and many other) words are also 'morphemes'.

Kempler Cohen's book consists of 96 pages of text and references, and two appendices, the first ('Lexical evidence for the uniform-root thesis') being the longer of the two, at 133 pages, and the second ('Early Austronesian morphemes reconstructed under the uniform-root thesis') being somewhat shorter, at 97 pages. The bulk of the book (230 pages) thus consists of a large body of data that is said to support the claims made.

I will comment on only one other point. Since he is concerned with AN languages as a whole, rather than focusing on a single language, Kempler Cohen also addresses issues of reconstruction. In a discussion of what he calls (again, idiosyncratically) 'The internal-reconstruction method' (1999:19), he has the following to say about Proto-Austronesian (pA): "This work therefore ventures to reconstruct proto-roots assignable to the primordial ancestor of pA . . . What is most astonishing is that this primordial ancestor probably arose as many as some

---

6 Examples are chosen from several sources: *pelay is from Dempwolff (1929), where it appears as *fole*, a form I have been able to find only in Samoan, where it is *fole* 'look pale, anemic; look ill, sick'; *kulay, *layas and *layu are glossed correctly, but *layaŋ is better glossed as 'hover', as of birds riding the wind'.

40,000 to 50,000 years ago. (This language presumably was just one of various primordial systems of sound-meaning associations devised in different locales around the planet and at different times in the early prehistory of humankind.)"

Since Kempler Cohen's references do not cite a single general treatment of historical linguistics, he can perhaps be forgiven when claiming time-depths for his reconstructions that no historical linguist studying any language family would take seriously. However, on the same page he follows up this bold statement with a series of remarks that make one wonder how much confidence he himself has in his results: "Because reconstructions of morphemes by the internal method are based on reconstructions of wordbases by the comparative method, the *phonetic composition* of such morphemes is inherently even less certain than the composition of their host wordbases. For this reason and others, I depict most internally-reconstructed morphemes generically, i.e. with each phoneme depicted by a sign ambiguously representing a group of alternative candidates for its position . . .

"Appreciably less tricky is the task of reconstructing by the internal method the *meaning* of a morpheme. In contrast with internal reconstruction of composition – a task which is hindered by phonetic variations among the relevant parts of the comparatively-reconstructed supporting wordbases – internal reconstruction of meaning often is actually facilitated by semantic variations among those wordbases, because additional glosses may provide better clues to the core meaning. And notwithstanding any variations, a common semantic denominator can invariably be ascertained."

Consistent with much of what is found in this book, the reader is left to wonder what it means to say that a meaning can "invariably be ascertained" in association with a morpheme (= root) of uncertain phonetic shape.

2 Issues relating to the root in Austronesian languages

The following terminology will be used in discussing issues relating to the root in AN languages:

Morpheme Isolate: A morpheme isolate is a morpheme found in a single language, or in a small number of languages that are too closely related to justify assignment to one of the proto-languages recognized here. A morpheme isolate contrasts with a cognate set, which is assigned to one or another proto-language. Each of these is etymologically independent—the attested forms by themselves and the cognate sets by the reconstructions to which they are assigned.

Root Isolate: A root isolate is a root that has no known variants, such as *-buŋ-$_3$ 'roof ridge', or *kep-$_2$ 'to seize, grasp, embrace.'[7]

Root Family: A root family is a collection of roots that show greater than chance similarity in shape and meaning. Some root families contain only two known roots, as *-keR/teR 'stiff, rigid, as a corpse', others contain three, as *-gaw/ŋaw/taw 'confused, disoriented, lost', and others contain many, as with *-ku/kuC/kug/kuk/kul/kuŋ/kuq, all meaning 'bent, curved' and the like, to which we might also add *-luŋ 'bend, curve'. Members of the same root family, which can be called 'root variants', or 'root doublets', generally differ in a single segment (for onomatopoetic roots this is most often the vowel); if they differ in more than one segment, as with *-cak 'the sound of walking in sticky mud' and *-tek 'mud, muddy' their status as members of the same root family is questionable (in this case *-cak is arguably onomatopoetic, and its similarity to *-tek may be due to chance).

In some cases a note has been added to an established root citing variant forms that occur in fewer than four EIMs, as with *-mul 'to hold s.t. in the mouth, suck on s.t.' (supported by six examples), but also Ayta Abellen *amal-amal* 'to suck on, as candy', Casiguran Dumagat *ʔəmél* 'to suck on s.t., as candy', and *məlməl* 'to stuff food in one's mouth', which are not yet sufficient to propose a separate root *-mel.

---

[7] By variants I mean independent roots of similar shape and meaning, such as *-bet/but 'buttocks', or *-kad/ked/kud 'to prop, support', rather than alternative forms of the same root which may or may not have contained a medial laryngeal, as with *-pit (or *-piqit?) 'press, squeeze together; narrow'.

Simple Morpheme: A simple morpheme is one that contains no smaller recurrent sound-meaning association within it, as with PAN *laŋiC 'sky, heaven', or PAN *puluq 'ten'.

Complex Morpheme: A complex morpheme is one that contains a phonestheme, or smaller sound-meaning association within it, which in the present study is the -CVC root of AN languages. Over 5,400 examples of such morphemes are given below.

Word Family: A word family is a collection of two or more morphemes that appear to be variants of the same form. It subsumes two types of variation. The first of these are doublets, which are simple morphemes that vary in one, two, or occasionally more than two features, as with PMP *tiduR/tuduR 'to sleep', PMP *kabu/kapuR 'kapok tree', or PWMP *betiq/beties/bities/butiqes 'calf of the leg'. The second type is found in complex morphemes, which are the focus of the present study.

Despite a fairly lengthy discussion of issues relating to the Austronesian root in Blust (1988b), there are a number of topics that were not discussed, and these will be treated here. The first of these topics is 'Finding the root'.

2.1 Finding the root

The first thing to note with regard to finding the root is that root-searching is more challenging than cognate-searching. The reason for this difference is that cognate morphemes normally show recurrent sound correspondences in all segments, and given the conventional alphabetic organization of information in the practice of lexicography (at least in languages that use the Roman script), comparing the initial syllable in words across a range of languages is far easier and less time-consuming than comparing the final syllable, which may be found in morphemes that begin with any unpredictable segment.

As will be noted at greater length below, the difficulty of identifying roots increases in languages that have lost final consonants, since the common historical process of 'erosion from the right' reduces the information load of final syllables. Brandstetter famously excluded the Oceanic languages from his treatment of comparative issues concerning Austronesian (his 'Indonesian') languages. While this was based on considerations other than the visibility of the root, problems in recognizing roots in many of these languages would have forced him to avoid using them extensively in any case. Nonetheless, among the 23 languages that he employed in searching for the root he included Bare'e, which has also lost all final consonants, although he cites it very sparingly.[8]

Even where all segments in a -CVC root have been preserved, roots may sometimes be overlooked because of metathesis in medial clusters, a process that is especially common in Philippine languages. The possible role of metathesis in inserting a glottal stop within a -CVC root will be examined in greater detail below, but metathesis also affects other consonants, and these often have a greater effect than the glottal stop in masking a valid root that has been altered by segmental transposition. Among many possible examples are the following, where the probable root that is revealed by comparative evidence is disguised by an intrusive consonant between the onset and nucleus of the final syllable.

(1)

Casiguran Dumagat
takləb 'cover of a pot, lid of a box, etc.' (root: *-keb 'lid, cover; to cover'?) (+M)
toklód 'pole used to hold open a window (root: *-kud 'walking stick; prop; support'?) (+M)
tukláb 'to pry open' (root: *-kab 'to open, uncover'?) (+M)

Keley-i
gəkyəm 'to grasp s.t. in a closed fist' (root: *-kem 'to clench, cover, grasp'?)
hiknul 'elbow' (root: *-kul 'to curl, bend'?)
siklup 'to close s.t.' (root: *-kup 'to enclose, cover'?)

Pangasinan
pəlnák '(of the sun) to rise' (root: *-lak 'to shine'?)
saklán 'to ride a horse astride' (root: *-kaŋ 'to spread apart, as the legs'?)
tikləb 'to be upside-down, lie on the stomach' (root: *-keb 'to lie face down, be prone'?).

---

8 The main languages used by Brandstetter (1916:9–10) were 1. Tagalog, 2. 'Bisaya' (= Cebuano), 3. Ilokano, 4. Sangir, 5. Tontemboan, 6. Bulu, 7. Bare'e, 8. Buginese, 9. Makassarese, 10. Balinese, 11. Madurese, 12. Javanese (= Old Javanese), 13. Sundanese, 15. 'Dayak' (= Ngaju Dayak), 16. Minangkabau, 17. Karo Batak, 18. Toba Batak, 19. Gayo, 20. Acehnese, 21. Mentawai, 22. Malay, and 23. Malagasy. He adds that "In a few cases some other languages besides these will also be used."

Kapampangan
balúktut 'bent, curled up' (root: *-kuC 'hunched over, bent'?)
kíldap 'lightning' (root: *-lap 'to flash, sparkle'?)
lakbáŋ 'pace, stride, step' (root: *-kaŋ 'to spread apart, as the legs'?)

Bikol
haklás 'to detach, peel off (as paper, decorations)' (root: *-kas 'to loosen, undo, untie')
haklóp 'to cover a frame, as of a kite or lantern' (root: *-kup 'to enclose, cover')
hulpós 'to slip or slide off, as a ring from the finger' (root: *-lus 'to slip off, slide down')

Hanunóo
pakpút 'application of sticky substance to a wound' (root: *-keC 'adhesive, sticky'?)
suklúb 'covering, top, as of a container' (root: *-kub 'a cover; to cover'?)
tikláŋ 'step, pace' (root: *-kaŋ 'to spread apart, as the legs'?)

Cebuano Bisayan
búnlut 'pull with force, as hair' (root: *-NuC 'to pull out, uproot'?),
dáplay 'to hang down loosely over an edge' (root: *-pay 'to drape over, hang down'?)
káblit 'to move s.t. by curling the fingers' (root: *-bit 'to hook, grasp with fingers'?)

In some cases both an original and a metathesized version exist, as with Cebuano Bisayan *úlyap ~ úylap* 'flare up, burst into flames' (root: *-lap 'to flash, sparkle'), supporting the inference that forms such as the sample of 21 listed above contain a root that has been disguised by secondary change. Nonetheless, to avoid unnecessary controversy, I have generally excluded such forms as evidence for roots unless an unmetathesized alternative also is reported. Intriguingly, no examples of onomatopoetic roots have been found so far with metathesis of the segments in a derived medial heterorganic consonant cluster, presumably because their imitative force would then be compromised.

Where there is a strong preference for disyllabic word bases of the shape CVCVC or CVNCVC, as in many Austronesian languages, systematic searching of base morphemes that end with a given -CVC is a feasible strategy in searching for the root, as the possible canonical shapes are fairly limited, and so allow a relatively quick check of possibilities. In my earlier study of the root this was done for Malay words ending with *-pit*, where the search included *apit, ampit, capit, campit, cəpit, cəmpit, cipit, cimpit, cupit, cumpit,* etc., and it was found that 19 of the 39 disyllables in Wilkinson (1959) which terminate with this sequence mean something like 'the approximation of two surfaces in a pinching movement', as *(h)apit* 'pressure between two disconnected surfaces', *capit* 'pincers', *men-cəpit* 'to nip', *dəmpit* 'pressed together, in contact', etc. (Blust 1988b:18–19).[9] However, quite apart from the fact that a mechanical procedure of this kind misses longer base entries, this is a single root in a single language. When the task is expanded to include several hundred roots and a similar number of languages, it becomes impractical to systematically search all available sources for possible roots. In dictionaries that have a reverse index searching by meaning can sometimes accelerate the process of identifying roots, but is likely to miss those that differ in semantic details, as with reflexes of *-bun-$_1$, which refer to dew, mist, fog, drizzling rain or blurry/dim vision (the assumed connection being the difficulty of seeing in foggy or misty weather, or when one's eyesight is poor). The result is that much of what happens in the data collection process is random, and hence is likely to miss relevant examples.

Although most roots that have been added since Blust (1988b) were discovered through random searching, in a few cases where I already had three examples of a root candidate and hoped to find a fourth to convert the candidate to an established root, I systematically searched a number of languages for the relevant -CVC sequence. This was easiest where a reverse index was provided, and the search proceeded from the translation gloss, rather than from the vocabulary of the target language. An example that failed this test (or has failed so far) is *-deŋ 'to extinguish a fire', attested in the cognate sets assigned to PAN *padeŋ, PWMP *pideŋ 'to extinguish, douse a fire', and in Binukid *pedeŋ ~ pereŋ* 'to extinguish a fire'. In other cases, a root candidate was successfully converted to an established root by targeted searching. This happened with *-kul 'snail', attested for some time in just Bintulu *bəkul*, Wolio *biku*, and Pangasinan *bisokól* 'snail' before Keley-i *batsikul ~ basikul* 'snail' was discovered, and also with *-huR 'to mix, stir', where systematic searching turned up Palawano *lehug* 'to mix, mixed' as a fourth example of this root.[10] However, the task of

---

9 The sequence bV(N)pV(C) is rare in Malay, and provided no examples.
10 At first glance Old Javanese *kul* 'a certain mollusk (mussel, oyster, clam?)', Javanese *kul* 'a variety of snail', and Balinese *ka-kul* 'snail, slug' appeared to provide a fourth etymologically independent example of *-kul 'snail', but closer attention to the comparative data shows that these forms reflect *kuhul 'edible snail sp.' (cf. Blust and Trussel 2020, sub PWMP *kuhul 'edible snail sp.'), and unless the well-established -CVC template for Austronesian roots is abandoned they cannot be considered related to the final syllable of Bintulu *bəkul*, etc.—a point to which I return below.

finding examples in this way is very much like searching for the proverbial needle in the haystack – it is time-consuming, tiring, often frustrating, and in the end a matter of luck.

In addition, one of the things that a researcher who looks for roots in AN languages will come to notice rather quickly is that not all languages are equally helpful in finding them. No attempt has been made to tabulate the citation frequency of languages in this study, but it is fair to say that most phonologically conservative languages of Taiwan, the Philippines and western Indonesia, together with some in eastern Indonesia, yield a relative abundance of roots without great effort in searching, while others prove to be far less productive. In particular, languages that have reduced final consonant contrasts, as many of those in Sulawesi, or lost original final consonants, as is true of the great majority of the 460 or so languages in the Oceanic subgroup, prove far less helpful in the search for roots than their relatives further west and north. As already noted, Brandstetter worked only with what he called "Indonesian" languages, excluding Oceanic languages from his research. However, even some languages that preserve the original -CVC of most word bases seem to have fewer submorphemic sound-meaning correlations than others of similar canonical shape. Chamorro and Palauan, for example, have yielded very few examples of roots in the present study, while Philippine and western Indonesian languages such as Ilokano, Maranao, Iban or Malay, or eastern Indonesian languages such as Manggarai, have yielded many more.

In short, then, I do not pretend that the present collection of data is complete, either in the number of roots identified or the number of EIMs that exemplify them. Indeed, picking up a dictionary or wordlist of any language that has not already been thoroughly searched quickly yields new examples, as any curious investigator can readily discover. Rather, what I offer is a representative collection of examples, most of which must be interpreted as final -CVC syllables, as other analyses (such as the -VCV- analysis of Nothofer 1990, 1991) simply do not work. As such, my hope is that it will be seen as a foundation for further research rather than as a *fait accompli*, for much still remains to be done.

2.2 Root ambiguity

Because they are short, and often belong to root families, many roots that appear in morpheme isolates which exhibit phonemic mergers can be assigned to more than one variant. For example, like most other Bisayan languages, Waray-Waray has merged PAN *e (schwa) and *u, making it impossible to determine whether a form such as *dasók* 'to keep or insert s.t. haphazardly or hastily' should be assigned to *-sek-$_2$ 'to insert, stick into a soft surface', or to *-suk 'to insert, penetrate, enter'. In all such cases I have assigned a phonologically ambiguous root to the variant with the larger number of unambiguous attestations, since this is more likely to be its source than its less robustly attested competitor. Since *-suk is justified by 35 EIMs that provide unambiguous support for the vowel, as opposed to 23 for *-sek-$_2$, Waray-Waray *dasók* is assigned to the first of these variants.

While this practice creates no issues in terms of the quality of evidence for a root, since both *-sek-$_2$ and *-suk are supported by far more than the required four EIMs, it can be problematic in cases where acceptance of a root variant is dependent on ambiguous data. For example, PAN *pacek 'to drive in (a nail, post)', Malay *cacak* 'planting upright, embedding' unambiguously support an underrepresented root variant *-cek-$_2$ 'to drive in by force, as a post' which could reach the required minimum of four EIMs if Cebuano *úgsuk* and *úsuk* 'to drive stakes into the ground' were included as support. However, since these Cebuano forms can equally well be assigned to the previously mentioned and far more robustly attested *-sek-$_2$ or *-suk, they cannot be used here to add another root variant to the dictionary, and *-cek therefore remains in the category of 'potential roots' awaiting further confirmation.

The most extreme cases of root ambiguity are found in languages that have lost final consonants. It is impossible, for example, to know whether -*la* with reference to shining light, reflects *-laC 'to shine; flickering or flashing light', *-lak 'to shine', *-laŋ 'to glitter, flash', or *-lap 'to flash, sparkle'. Although all of these root variants are well-supported by examples that preserve the final consonant (*-laC with 13, *-lak with 16, *-laŋ with 9, and *-lap with 35), as in other cases where an arbitrary choice must be made regarding the assignment of a complex morpheme to a root variant, I have favored the most robustly attested variant as the most likely source. In the present case, then, I have put Label *hil* 'lightning; to flash, of lightning', Dobel *ŋela* 'lightning', Tolai *pala* 'to flash', Wolio *tila* 'light, radiation, glare, mirror; to shine upon, irradiate, illuminate', and Lau *tala* 'to shine, of sun or moon' under *-lap, although they could as easily have been put under any of the other variants (but cp. Lau *talaf-i* 'to shine on, enlighten, lighten up', where the assignment for this language is justified by the thematic consonant that appears under suffixation).

Without these controls on inference it would be possible to propose additional roots that are dependent on a generous interpretation of ambiguity. If Wolio *siru* 'ladle, spoon', Erai *hahuru* 'spoon' (both of which have lost final consonants) were assigned to *-du rather than the equally possible and better-attested *-duk 'ladle, spoon', for example, it would produce the required four examples needed to justify a root doublet, or more properly, 'disjunct' (Blust

1970:112–13), but only by accepting two that are ambiguous (eventually the random discovery of Central Tagbanwa *karo* 'ladle', which reflects *-k without change, allowed the inclusion of *-du in the same root family as *-duk). This problem is particularly serious in distinguishing *c*-initial roots from *s*-initial roots, since only about two dozen languages reaching from Sumatra to Sulawesi distinguish these phonemes (Blust 2013:570–74). To sum up, then, in the interest of maintaining high standards for the recognition of roots, wherever an arbitrary assignment of this kind would allow positing a new root I have instead assigned the morpheme in question to a root that has been established on the basis of unambiguous evidence.[11]

A somewhat different problem is that of contradictory indications. PAN *t and *C are distinguished by most Formosan languages, but merged as *t in PMP. Some morphemes that distinguish these phonemes in a particular root point to *C, as with *balalaCuk 'woodpecker' (cf. Saisiyat, Taai dialect *baLasok* 'woodpecker', Paiwan *balatsuk* 'Formosan barbet', both indicating *-Cuk), but others point to *-t, as with PAN *tuktuk 'to knock, pound, beat' (cf. Puyuma *tuktuk* 'to hit with a hammer', indicating *-tuk). It is hard to know what to do with such material, which is clearly contradictory, yet apparently related. To avoid massive duplication (since all non-Formosan forms could reflect either *-Cuk or *-tuk), I have put all of this material under *-Cuk, with a notation that Formosan languages vary in sometimes merging this with *t, and sometimes maintaining the distinction.

2.3 Control of chance

The most critical question in research on the root in AN languages probably is "When do we know that what we have found is valid, rather than a product of chance?" Although I have drawn on only slightly over 200 languages and dialects here, there are over 1,200 AN languages (Eberhard, Simons and Fennig 2021), so it is likely that some examples of corresponding -CVC sequences in forms of the same or similar meaning will arise by chance. How many examples of this kind can be expected solely through the operation of chance is difficult to determine. Blust (1988b) adopted two strategies to minimize the probability that an apparent root is a product of random agreements when lexical material

---

[11] Note, however, that where there is no competition for assignment to a root variant, ambiguous evidence needed to reach the minimum of four etymologically independent attestations is accepted, as with Wolio *biku* for *-kul 'snail'.

can be drawn from hundreds of languages, and these strategies are continued here. The first is to require that all roots be attested in four or more EIMs. The second is to require that a root receive significant crosslinguistic support, meaning in essence that it can be assigned to one or more well-established proto-languages.

There is nothing magical about the number four. As noted earlier, the roots in Blust (1988b) varied from *-pit 'press, squeeze together; narrow', with 48 supporting cases, to 15 roots with just 4. Needless to say, a root like *-pit, which has been identified with the same or very similar meaning in four dozen cognate sets or morpheme isolates, is more secure than one like *-ɲis 'cruel', which has so far been found in just four. Some 109 of the roots in Blust (1988b), or nearly half, are found in 10 or more EIMs, and 24, or over 10% are found in 20 or more. The numbers in the present study are far higher, making it unlikely that they are products of either chance or alternative analyses. Nevertheless, there are root candidates that have so far been identified in only two or three non-cognate morphemes, and it is an open question whether these are valid roots that will be firmly established by the discovery of additional data, of whether they are products of chance. The most questionable examples in the present dictionary are those that are attested in just four EIMs. Although a few of these may be due to chance, this is unlikely as a general explanation, as more than two or three of the roots proposed here would not be expected to arise by chance four or more times in a collection of related languages, even if these number in the hundreds. Add to this the fact that the 406 roots proposed in this work are supported by 5,404 EIMs, and it follows that the average root contains about 13.3 supporting examples.

Another consideration in minimizing chance as an alternative explanation of the data is that a candidate which is assigned to a proto-language is more likely to be valid than one claimed for a single language. This is true for at least two reasons. First, as already noted, many attested languages have undergone phonemic mergers, including merger with zero through the loss of final consonants. In the latter case, one might doubt that the final -CV in one language is part of the same root as the final -CVC in other, as in asking whether Wolio *biku* 'snail' contains a reflex of the root *-kul noted in 2.1, or fortuitously resembles the other forms with which it has been compared. As a general result of this condition few Oceanic languages have been used in this study, since 80–90% of the 460-odd members of this group have lost original final consonants, and a number have also lost the vowel that immediately preceded it, leaving only the initial consonant of an original -CVC.

Second, studies of submorphemic sound-meaning sequences in individual AN languages such as Malay/Bahasa Indonesia, or Tagalog have shown a tendency to treat semantics less strictly than is usually true of the -CVC root in the work of either Brandstetter (1916) or Blust (1988b). While the meanings of cognate morphemes clearly may vary, and some allowance must also be made for this in identifying-CVC roots, it is methodologically advisable to insist that meanings be identical or similar enough to ensure interpersonal agreement. Again, in the interest of maintaining high standards for the recognition of roots, I have adhered to fairly strict requirements on semantic matches. The most noteworthy departure from this practice has been with onomatopoetic roots, which often refer both to a natural sound and to the action or instrument used to produce it. The claim that, e.g. PAN *balalaCuk 'woodpecker', PPH *suntuk 'to punch', Puyuma (Tamalakaw) *mu-palTuk* 'to explode', and Manggarai *gatuk* 'to chatter, of the teeth' (among others) share a root PAN *-Cuk, PMP *-tuk 'to knock, pound, beat; the sound of these actions', for example, is based on the observation that the action of knocking, pounding or beating normally produces a sound that is associated with it, and in any given morpheme the root may represent either the action or the sound (or both), a generalization of semantics that applies to several onomatopoetic roots, and thereby strengthens the claim that a similar process is operative in all of them.

2.4 Etymological independence/control of duplication

As already stressed in Blust (1988b), it is essential that the number of morphemes in which a root is attested be distinguished from the number of cognate sets that contain it, since ignoring this condition would artificially inflate the number of confirming cases. To choose one of many possible examples, Bunun *dumdum* 'dark, overcast, gloomy (as the sky on a cloudy day)', Itbayaten *rəmdəm* 'cloud', Tagalog *limlím* 'impending darkness in late afternoon', Kelabit *dədhəm* 'dark; darkness', Muna *rondo* 'dark, be night', Chamorro *homhom* 'dark, dim, obscure, dusky, gloomy', Wuvulu *xoxo* 'dark', and Woleaian *rosh* 'night, darkness; dark, black, obscure' all appear to contain the root *-dem 'dark, overcast', but cannot be counted as eight supporting cases, since they all reflect PAN *demdem 'gloom, darkness; dark, overcast, gloomy' (cf. Blust and Trussel 2020 where 39 reflexes of this reconstructed form are cited). Forms that are cognate thus provide no more support for the reality of a root than morpheme isolates like Balinese *surəm* 'obscure, dim', Puyuma *ʔudədəm* 'black', Isneg *xídam* 'evening, from about 5 P.M. until darkness sets in', or many others that—so far as is presently known—have no cognates in other languages.

The treatment of cognate sets as single forms reduces the number of potential supporting cases for the reality of a root, but in addition to avoiding the methodological error of duplicating evidence, it has two other advantages. First, as already noted, proto-forms are more likely to retain phonological distinctions that have been lost through merger in attested languages, and they therefore help to avoid ambiguity. Second, the assignment of roots to reconstructed morphemes provides evidence that some roots must have a long history in the AN language family.

Another factor that may lead to inflation of evidence is unrecognized morphology. Where the morphology is still productive this is relatively easy to control, as most lexicographers list the vocabulary of AN languages under the base, with derivatives arranged under it. However, if the morphology is rare, poorly understood, or fossilized, morphologically complex words may be mistaken as distinct morphemes. This is most often encountered in words that contain more than two syllables, and end with the same -CVCVC, as with Bikol *kitkilát* 'lightning', which is counted as a partial reduplication of PAN *likaC, PMP *kilat 'lightning', and hence not etymologically independent, or Long Moh Kenyah *pelakut* 'to bend over, of ripe rice', next to *lakut* 'bent over' (cp. e.g. *tulat* 'portion, share', but *pe-tulat* 'to divide', or *katu*, *pe-katu* 'to send', which show that *pe-* is a verb prefix, but it is unclear whether *pelakut*, which is listed separately and cross-referenced to *lakut*, is bimorphemic or monomorphemic). A similar problem arises with apparent CVCCVC reduplications with -VC- infixes which are given as lexical entries in standard dictionaries. Bikol, for example, appears to have had infixes *-ar- and *-aR- within the relatively recent past, as seen in examples such as *parikpík* 'fin, flipper' (PAN *pikpik 'sound of patting or tapping'), *tariktík* 'woodpecker' (Bikol *tiktík* 'to dislodge s.t. by tapping a container gently with the fingers'), *pagakpák* 'the sound of flapping wings' (Bikol *pakpák* 'wing, as of a bird'), or *pagukpók* 'a knocking sound' (Bikol *pukpók* 'to heat, hit or strike with a stick'). In general, I have treated these as etymologically independent only where the CVCCVC reduplication and its infixed counterpart show marked differences in meaning, as with *parikpík*, next to *pikpik. As just seen, the most difficult cases are ones in which the morphology is fossilized, and hence perhaps rare, and it is possible that in some cases I have mistakenly included affixed forms as though they are independent morphemes.

Somewhat different is control of duplication as a result of irregular sound changes. To illustrate, since PAN *u normally did not change in Kankanaey, should Kankanaey *sokúd* 'to lean upon (a staff, cane or support)' be regarded as an EIM in support of the root *-kud 'cane, staff, walking stick', or as a phonologically irregular member of the cognate set assigned to PAN *sukud 'walking stick, cane, staff'? If the former option is chosen it increases the number of *apparent* EIMs supporting *-kud by assuming that *sokúd* was an independent lexical innovation that shares a common root with reflexes of PAN *sukud, rather than a reflex of *sukud with irregular change. On the other hand, if the latter option is chosen it reduces the amount of potential support by treating phonologically incompatible forms as though they are etymologically equivalent. In general, I have treated such forms as lexical innovations, and hence as etymologically independent evidence for the root in question. While the decisions I have made in such cases may be controversial, issues of unrecognized morpheme boundaries, homophonous complex morphemes, and irregular sound change affect a fairly small percentage of the total database, leaving the broad conclusions largely intact.

Low level subgroups or lexical distributions that are geographically restricted present another kind of problem. Whereas cognate sets are normally combined under a single proto-form, those that are known only among closely related languages or languages that are geographically contiguous cannot easily be represented by a reconstruction for which an appropriate label can be found. The present study recognizes PAN, PMP, PWMP, PPH, PCEMP, POC, and occasionally other proto-languages, but in a comparison such as Buginese *babaŋ*, Balaesang, Wolio *bamba* 'door', all of which are spoken in Sulawesi, and which appear to reflect *baŋban (root: *-baŋ 'door'), it is unclear what proto-language this form should be assigned to. Similarly, Ida'an Begak *limbaw* 'shallow', Kenyah *libaw* 'a ford; shallow' seem clearly to be cognate, but again an appropriate proto-language is elusive. In this and similar cases in other areas, I have generally chosen just one language to represent the collection of languages that share a cognate set of undetermined antiquity, as with members of the putative Greater North Borneo group (Blust 2010).

Still another problem relating to etymological independence is seen with homophonous proto-forms that are semantically similar, but not identical, as with *-sek$_1$ 'to cram, crowd', and *-sek$_2$ 'to insert, stick into a soft surface'. While it is conceivable that these are a single root, the clustering of supporting forms around one meaning or the other is sufficiently clear that there seems to be adequate reason for distinguishing them.

Finally, there is sometimes a question of whether morphemes are etymologically fully independent, or only partially independent. For example, while Aklanon *púsdak* 'to stomp, bang one's feet', and Malay *jəjak* 'to trample on' are excellent examples of EIMs that appear to contain the root *-zak to step, tread, trample', Tagalog *buklát* 'open, opened', Hiligaynon *múklat* 'to open one's eyes in waking', Aklanon *múdlat* 'to open up the eyes very wide', Cebuano *búdlat* 'bulging eyes; for the eyes to bulge', all of which evidently contain the root *-lat 'to open the eyes wide', form a continuum which suggests that speakers may have simply reworked the same complex morpheme into varying shapes.

Whether to count all such examples as EIMs, to treat them all as one, or to find some compromise between these extremes is largely arbitrary. Fortunately, such cases are rare, and regardless of what decision is made, it will have very little effect on the overall picture.

2.5 Root and doublet

A basic question that might be raised in discussions of the root in AN languages is "What is the difference between doublets and morphemes that share a root?". In English and many other languages the term 'doublet' refers to words of the same origin in a language that has inherited one directly, and borrowed the other from a related language, as with English *shirt* (native) and *skirt* (early Scandinavian loan), or *ship* (native) and *skiff* (loan from Middle French). However, many AN languages have formally and semantically similar words, both or all of which appear to be native. An extensive examination of these appears in Blust (2011). Some common examples include PMP *tiduR and PAN *tuduR 'to sleep', PAN *beli and PMP *bili 'to buy', and PMP *kumaŋ PAN *qumaŋ hermit 'crab', where both members of each doublet pair co-existed in PMP. These examples and many other patterns of doubletting can clearly be distinguished from morphemes that share the same root, but in some cases the distinction is unclear. For example, do we treat PMP *lamuk and *ñamuk 'mosquito' as sharing a root *-muk 'mosquito', much like the seven forms cited under *-baw-$_2$ 'rat, mouse' in the present collection, or do we treat these two examples instead as doublets?

Since this question arises only where two or more semantically similar forms share a common -CVC, it concerns a relatively rare situation, as most words classified as doublets in Blust (2011) differ in the vowel of either of the last two syllables, or in some other way. In general, semantically similar forms which share a common -CVC were treated as doublets if only two variants were known, as PMP *taked/tiked 'heel', or *lamuk/ñamuk 'mosquito'. However, as the present work makes clear, this situation can change as more comparative research is undertaken. Blust (2011:439), for example, gives PMP *lamuk/ñamuk 'mosquito' as doublets, but since 12 EIMs that end with a reflex of *-muk, have now been identified with this or a very similar meaning, their similarity is now better considered a product of root-sharing, making them complex morphemes, which contain a submorphemic sound-meaning association, rather than doublets, which do not.

2.6 The shape/structure of the root

As already noted, under the heading 'Related matters', Blust (1988b) noted types of sound symbolism in AN languages that are distinct from the root, and that therefore lack its properties. These include single-consonant onsets such as *k-* or *-g* in many words that refer to rubbing, grating, scraping or scratching, or codas like *-l* in many others that mean 'blunt, dull', as well as configurations of segments and syllables that were described under the heading 'Gestalt symbolism'. However, as noted already, these types of sound symbolism are much less commonly encountered than the -CVC root.

The AN root as recognized by Brandstetter (1910) and Blust (1988b), has several generally accepted properties. First, it is a complete syllable rather than a syllable fragment as in the English phonesthemes *sn-* 'nasal area' (*sneer, sneeze, sniffle, snore, snot, snout*, etc.), or *-ash* 'violent collision' (*bash, clash, crash, dash, smash*, etc.). Second, it is almost always -CVC, and hence is a single closed syllable. Finally, it is almost always the last syllable in a morpheme. Each of these assumptions has been challenged by one writer or another, but the bulk of reliable evidence strongly supports them, with certain qualifications to be noted below.

The second property of the AN root, namely that it almost always is a closed syllable, is perhaps the one that is most open to challenge. Blust (1988b) recognized *-bu 'dust' as a -CV root in eleven EIMs, with a question as to whether it might have been *-buh, given the Amis form *alafoh* 'dust blown by the wind'. The latter interpretation is now adopted (as *-bux), suggesting that even roots that have the shape -CV in nearly all languages were originally -CVC. However, the data collected in the present study includes five new -CV roots, *-ka 'to split', *-ni 'to hide, conceal oneself or s.t.', *-ŋa 'to gape, open the mouth wide', *-pi$_1$ 'dream', and *-pi$_2$ 'to fold'. Three of these have variant forms with final consonants (*-kaq 'to split', *-ŋab, *-ŋaŋ, *-ŋap, *-ŋaq, all meaning 'to gape, open the mouth wide', and *-piq 'to fold'). The variants that lack a final consonant are supported by unambiguous evidence that distinguishes them from other members of the same root family, and so far as can presently be determined, these are likely to remain in violation of the -CVC pattern, since there is no evidence that any of these roots contained *-h or some other laryngeal consonant that disappeared in all languages outside Taiwan (Blust 2014:233).

In addition, both Nothofer (1990, 1991) and Kempler Cohen (1999) claim to have found semantically consistent CVC- sequences in word-initial position, but there is very little evidence that any of the -CVC roots proposed in Blust (1988b) or the present work can appear word-initially. The sole example cited to date is Malay *ñaman* 'tasty, savory', Keley-i *namin* 'to taste a little', which appears to contain the same root found in PAN *ñamñam 'tasty, delicious', Javanese *kəñam* 'taste in the mouth', Matu Melanau *kuñam* 'taste', etc. As already seen, finding -CVC roots is far more difficult than finding the same phoneme sequence in initial position, and if roots were a part of word structure in initial position, more than one of them would surely have been noticed by now.[12]

Blust (1988b) noted that only onomatopoetic roots can occur in isolation, as with PMP *bak 'clap! smack!', PMP *tuŋ 'deep resounding sound', PMP *gur 'purring or grunting sound', or Malay *tok* 'a dull knock'. In these cases the root is a simple morpheme, as it is not embedded as a submorphemic sound-meaning association in a larger phoneme string. However, where a root is found in a larger morpheme, but never in isolation, it is not a morpheme, as the *-beŋ* in PAN *beŋbeŋ 'blocked, as by a wall or curtain', PMP *embeŋ 'a dam; to dam a stream', Itbayaten *rivəŋ* 'shield, lee, windbreak', Madurese *tebbeŋ* 'closed off with a wall', Vitu *zobo* 'blocked (or road, pipe)', or many other non-onomatopoetic roots in Blust (1988b) or the present collection.

The topic of root variation was discussed at some length in Blust (1988b), where it was noted that doubleting occurs with roots, just as it occurs with base morphemes. Some roots show limited variation, as with *-beŋ 'to block, stop, dam', next to *-peŋ 'to plug up, dam; cover', while others express so many phonemic variants that one must wonder how they differ semantically, as with the runaway series *-ku 'bend, curve', *-kuC 'hunched over, bent', *-kug 'curl, curve', *-kuk 'bent, crooked', *-kul 'to curl, bend', *-kuŋ-₁ 'to bend, curve', and *-kuq 'to bend, curve'. Most root variation appears to be random in that the varying segments have no effect on the meaning of the root. Thus *-beŋ and *-peŋ occur in morphemes that do not differ semantically in any systematic way, and this is equally true of the entire series that begins with *ku- with reference to bending, curving and the like. However, with onomatopoetic roots the matter is different.

In collecting data for over 200 roots Blust (1988b) noted a pattern of size/sound symbolism associated with the voicing of stops and the quality of vowels, and this continues to be supported in the expanded database presented here. In general, a stop coda represents a sound in nature that terminates abruptly, while one ending in a nasal (usually the velar nasal) represents one

---

[12] Zorc (1990) has proposed a few additional examples in initial position, as PMP *taŋan, which he analyzes as *taŋ- + *-an, or *paRaw 'hoarse', which he analyzes as *paR- + *-aw. However, plausible examples of recurrence for *taŋ- in the meaning 'grasp', *paR- in the meanng 'hoarse' or an initial CVC- in the few other forms that he cites in this connection, do not emerge from the data in Blust and Trussel (2020), or any other source I am aware of.

that continues to resonate after the action that produced it has ceased. Thus, morphemes that contain the root *-bek 'dull muffled sound' commonly refer to sobbing, chopping, thudding and the like, while those that contain the root *-beŋ 'dull resounding sound' commonly refer to buzzing, humming, groaning, or the sound of drums. While this contrast in the imitative value of stops vs. nasals is arguably universal (think of 'tick-tock' vs. 'ding-dong' in English), the onomatopoetic properties of AN roots extend beyond the 'abruptly terminating' vs. 'resonating' distinction into other qualities of sound. One of these qualities is expressed in the vowels. As noted in Blust (1988b:37ff), onomatopoetic root variants that differ in their vowel show systematic differences in the type of sound that is represented by /a/ (harsh, discordant), /e/ (dull, muffled), /i/ (high-pitched, shrill), or /u/ (deep, booming). Examples are seen in the series *-gak 'a raucous, throaty sound', *-gek 'a dull throaty sound', *-gik 'a shrill throaty sound', and *-guk 'a deep throaty sound', and parallel differences signalled by vowel quality occur in other root families, including *-pak 'to slap, clap', *-pek 'sound of breaking (as of wood)', *-pik/ 'to pat, slap lightly', *-puk 'to thud, snap, crack, break', and *-tak 'the sound of cracking, splitting, knocking', *-tek 'a clicking or light knocking sound', *-tik 'a ticking sound', or *-Cuk 'to knock, pound, beat'. Particular examples from attested languages often make these distinctions of sound quality quite clear, as reflexes of *-gak frequently refer to cawing, cackling and the like, reflexes of *-gek to choking, gulping, or sobbing, reflexes of *-gik to giggling, squeaking and similar high-pitched sounds, and those of *-guk to gargling, snoring or grunting, while reflexes of *pakpak commonly refer to the clapping of a bird's wings when taking flight, but *pikpik refers to the flapping of a fish's fins.

Another parameter of onomatopoeia in root variation is seen with the voicing of stops. As noted in Blust (1988b:42), the choice of a voiced stop either as onset or as coda in an onomatopoetic root "frequently designates a **louder** sound—one typically (though not always) produced by a more massive inanimate object or a larger animate being." Examples with the root *-bak include the sound of a stone or book falling (Paiwan *tsebak*), of banana stems clashing against one another in a strong wind (Kankanaey), of pounding foodstuffs (Bikol), and of hammering (Tiruray) or stamping (Yamdena). By contrast, examples with the root *-pak include the sound of slapping someone in the face (Bontok, Kankanaey), of a door slamming or a plate breaking (Ilokano), of wings flapping or hands clapping in applause (Tagalog), or the sound of water lapping on a beach (Malay). Collins (1979:387) has claimed instead that "the voiced stop is associated with events involving greater resonance," and Nothofer (1990:134) agrees: "Collins' analysis of the meaning of initial stops seems to be more correct than that suggested by Blust . . . Blust's own examples support

Collins' interpretation: for example, Yamdena *sambak* 'stamp on the ground with the feet or something else', Bontok *dospak* 'slap someone's face with the open palm of one's hand.'" However, a careful consideration of the onomatopoetic roots in Blust (1988b) and the present corpus lends no support to this claim. As the data given here attests repeatedly, sounds expressing greater resonance are represented by terminal nasals, almost never by stops, whether voiced or voiceless, and the contrast in meanings associated with morphemes that contain a voiced stop rather than a voiceless stop, as *-bak vs. *-pak, is consistently associated with a larger sound-producing object and/or, a louder sound, as a stone or book falling on a hard surface, as against people clapping, or a bird flapping its wings. Compare, for example, the forms in column A against those in column B:

(A) Final stop   (B) Final nasal

*-bag 'the sound of a heavy smack'
*-bak$_2$ 'the sound of a heavy smack'
*-bek$_1$ 'dull, muffled sound'   *-beŋ$_2$ 'a dull, resounding sound'
*-buk$_2$ 'to pound; a thud, heavy splash'
*-cak$_1$ 'the sound of walking in sticky
    mud'
*-Ceg 'to hit, beat'
*-Cug 'to pound, hit'
*-Cuk 'to knock, pound, beat'   *-Cuŋ 'a deep resounding sound'
*-dek$_1$ 'to hiccough, sob'
*-dek$_2$ 'pulverized, pounded fine'
*-gak$_2$ 'a raucous, throaty sound'
*-gek 'a dull throaty sound'   *-geŋ 'to hum, buzz'
*-gik 'a shrill throaty sound'
*-guk 'a deep throaty sound'   *-guŋ 'a deep resounding sound'
*-kak 'to cackle, laugh loudly'   *-kaŋ$_2$ 'to bark, croak'
*-kek 'to shriek, creak, cluck, chuckle'   *-keŋ$_2$ 'a hollow, resounding sound'
*-kik 'a shrill throaty sound'   *-kiŋ 'a clear ringing sound'
*-kiq 'a high-pitched vocal sound'
*-kuk$_2$ 'sound of a sob, cackle, etc.'   *-kuŋ$_2$ 'a deep resounding sound'
                                                *-leŋ 'a reverberating sound that
                                                    stuns a person'
                                                *-liŋ$_1$ 'a clear ringing sound'

*-ŋak 'to screech, howl'
*-ŋek 'to grunt, groan'   *-ŋeŋ 'to buzz, hum'
*-ŋik 'a shrill throaty sound'   *-ɲiŋ 'a shrill buzz or hum; to ring'

*-ŋuk 'a deep throaty sound'
*-ŋut 'to mumble, murmur, mutter'
*-pag 'to strike, beat'
*-pak-₁ 'to slap, clap'
*-peg 'to hit, beat'
*-pek 'the sound of breaking; powder'
*-pik 'to pat, slap lightly'
*-put 'to puff, blow hard'
*-riC 'the sound of ripping, etc.'

*-ŋuŋ 'deep buzzing or humming'

*-Raŋ-₂ 'clanging sound'
*-Reŋ 'to groan, moan, snore'
*-Riŋ 'to ring'
*-Ruŋ 'to roar, rumble'

*-tak-₂ 'the sound of cracking, splitting, etc.'
*-tek-₁ 'a clicking or light knocking sound'
*-tik-₂ 'a ticking sound'
*-tut 'flatulence'

*-taŋ 'a clanging sound'
*-teŋ-₁ 'to hum, drone'
*-tiŋ 'a clear ringing sound'

Several observations are worth making about this subset of roots. The first is that onomatopoetic roots that end with a stop are more common than those that end with a nasal (35 examples of the former, 20 of the latter). This difference presumably is because most sounds that these languages attempt to imitate are sounds that terminate abruptly. The second is that in 14 cases there are matching pairs of roots that are identical except that one ends in a (usually voiceless) stop, and the other in a nasal. The chief exception to this pattern is seen with roots that begin with a liquid (*l or *R), as there are six of these that end with a nasal, and no corresponding roots that end with a stop. The third observation that is of some note is that the final consonant in onomatopoetic roots is almost always a velar. For stops this is true in 30 of the 35 cases, and for nasals it is true of all 21 cases. The fourth observation worth making is that the initial consonant of onomatopoetic roots does not favor any particular class of stops (labial = 10 examples, palatal = 1 (*c-), alveolar/dental = 9, velar = 9), but all five nasals as the onset of an onomatopoetic root are velar. Fifth, as already noted, in onomatopoetic roots the value of the vowel generally signals the quality of the sound represented (a = harsh, e (schwa) = dull, muffled, i = high-pitched, u = low-pitched or deep).

The above generalizations hold up well against the data collected. In a few cases one might question whether the stop/nasal difference truly signals a difference of sounds that terminate abruptly vs. those that reverberate on for a moment longer, as with *-kak 'to cackle, laugh loudly' vs. *-kaŋ-₂ 'to bark, croak', both of which occur in morphemes that refer to various sounds produced by dogs, but in general *-kak is associated more often with the cackling of chickens and the like, while *-kaŋ-$_2$ more often signals the hooting of an owl or the noisy croaking of a bullfrog.

Non-onomatopoetic roots also vary widely in both vowels and consonants, but here the motivation and patterning have no obvious perceptual motivation. Why do roots that refer to radiant light show up in four known variants, *-laC, *-lak, *-laŋ, and *-law, all of which begin with *l-, but differ in the coda, while other non-onomatopoetic roots show up with variants that differ in their initial consonant or vowel, as with *-gaw/ŋaw/taw 'disoriented, lost', or *-get/git/gut 'angry'? In short, the patterns of variation in non-onomatopoetic roots are as puzzling as the patterns of variation in simple morphemes of the kind explored at some length in Blust (2011).

Finally, while most onomatopoetic roots end with either a stop or a nasal, a smaller subset ends with a sibilant or liquid. The examples currently available for the first of these categories are *-gis 'to scrape', *-kis-$_1$ 'to scratch, grate, scrape', *-pes 'empty, deflated', *-pis-$_1$ 'to deflate; empty', *-pus-$_2$ 'the sound of escaping air',*-ris-$_1$ 'a rustling sound', and *-rus-$_1$ 'a rustling sound', and those for the second are *-gur 'to purr', *-kur 'to coo; turtledove' *-ŋur 'a low-pitched sound', and *-sir 'a hissing sound'. Roots that end with a sibilant appear in general to represent natural sounds that have a scratchy, grating, or hissing quality, while in the few examples available those that end with a liquid describe the low rumbling of distant thunder, or the soothing sounds of purring or cooing. This is not always true with the roots *-pes or *-pis, which often refer to the reduction of swelling of a boil, to a rice husk without a grain, and the like where no sound is involved, but reflexes of *-pus usually have a more clearly onomatopoetic use, as in Malay dəmpus 'to puff in one's sleep', Malay dəpus 'the sound of a rush of air through a narrow opening, e.g. from bellows', or Javanese kəpus 'to blow on, exhale strongly'.

The other notable exception to his pattern is *-sir, which invariably appears in morphemes that refer to hissing or similar types of sound, as in Malay desir 'the hiss of water turning into steam', Kayan kesih 'the singing of water about to boil', Malay sir 'a hissing sound, such as that made by water dropping on red hot iron', or Makassarese tisiʔ 'to make a hissing sound with the mouth when eating hot foods, such as chili peppers'. In this particular root the sibilant onset seems to have a dominating influence on the type of sound represented.

One other template for the root that might be contemplated based on overlapping root shapes is one with only two phonemes, hence CV- or -VC This works particularly well with root variants that share the sequence -VC, such as *-leb and *-ñeb 'to sink, disappear under water', or *-dem 'dark; overcast', *-lem 'dark; obscure', and *-Cem 'dark, of color; a dark color', where each of these and a number of other root families are united by the recurrent partial *-VC. Other root variants differ in the final consonant, as with *-cik 'to fly out, splash, spatter', *-cit 'to squirt out', and might be said to point to a CV root. Still others differ only in the vowel, and so might be analyzed as containing a discontinuous root C . . . C, such as *-b . . . k in *-bek 'rotten, crumbling, pulverized', next to *-buk 'to decay, crumble; powder'. However, there are reasons for caution in adopting this type of approach.

First, with reference to the foregoing examples, if *-eb is a root meaning 'to sink', or *-em is a root meaning 'dark' or 'black', we must ask why the number of consonants that can precede these -VC sequences is so limited, despite relatively large numbers of EIMs that support them (12 for both *-leb and *-ñeb, 9 for *-Cem, 38 for *-dem, and 40 for *-lem). Second, as noted in Blust (1988b:37ff), onomatopoetic root variants that differ in their vowel show systematic differences in the type of sound that is represented, and it would therefore be awkward to describe cases such as *-gak 'a raucous, throaty sound', *-gek 'a dull throaty sound', *-gik 'a shrill throaty sound' and *-guk 'a deep throaty sound' as discontinuous roots *-g . . . k, since the quality of the vowel is an essential part of the meaning of these -CVC sequences. Third, root variation is not a novelty when we consider the need to recognize robust variation on the level of the morpheme, as in *ejuŋ/ijuŋ/ujuŋ 'nose', *tiduR/tuduR 'to sleep', and numerous other sets that have been described elsewhere (Blust 2011).

The most vexing issue relating to the shape of the root probably is the presence of pre- or postconsonantal glottal stops within a root in some Philippine languages, an issue that has so far been addressed only by Zorc (1990:180–81). Zorc framed this problem in terms of a reduction hypothesis, in which some roots were disyllables with a medial glottal stop that metathesized with the first consonant of a -CVC root, vs. an expansion hypothesis in which the glottal stop was added in the separate history of one or more languages. The appendix of roots in his paper lists only monosyllables or monosyllables with a possible CVqVC option, as *-pit 'to press, squeeze together; narrow', in order to account for examples such as Cebuano lipʔit 'put s.t. in between two flat surfaces', derived in Blust and Trussel (2020) from PMP *liqepit by schwa syncope and regular metathesis of glottal stop to postconsonantal position. Support for the claim that some of these -CVC roots were actually CVqVC is seen in examples like PWMP *piqit 'closed tightly, as the eyes', or PPH *sakaʔaŋ 'to walk or stand with legs wide apart' where there is little choice except to reconstruct a disyllabic form with a medial laryngeal.

A solution to determining the history of such forms is not proposed here, as it is part of a larger issue regarding the history of consonant clusters with glottal stop in Philippine languages that I hope to address at greater length in another publication. However, for the present it is useful to note that if some roots were disyllables such as *piqit, as several languages appear to indicate, it is surprising that all disyllablic roots contained a medial laryngeal, since no other medial consonant has been proposed in them. In languages that lack a glottal stop the normally -CVC root may sometimes appear with a doubled vowel, as in Kankanaey *liŋáʔət* 'to grate, produce a harsh sound', with apparent root *-ŋeC 'angry; to gnash the teeth'. In such cases, I assume that the double vowel was separated by a glottal stop at an earlier stage in its history.

A striking structural property of the root is that the ten most robustly attested roots (*-pit-$_1$, *-Cuk, *-kaŋ, *-keC, *-pak, *-tik, *-kuŋ-$_1$, *-kas-$_2$, *-puk-$_1$, and *-tas) begin with voiceless stops other than *q, and the most robustly attested nasal-initial roots begin with the velar nasal, as with *-ŋa, *-ŋit and *-ŋuC. At the other end of the frequency spectrum, only seven of the 29 least robustly attested roots begin with a voiceless stop (see Display 2 below). This shows only partial agreement with the phonological pattern for morphemes in Blust and Trussel (2020), where the three most frequent onset consonants for stops are *b- (1149), *t- (910), and *k- 865), with *p- (709) following both *s- (773), and *l (750), and the most frequent onset consonant for nasals is *m- (156), followed by *ŋ- (99), *n- (80), and *ñ- (26). The rarest consonantal onsets for roots recorded so far are *q- (1), and *h- (2).

The last thing to note is a pattern of semantic reversal using the vowels *e (schwa) and *a in a -CVC root, where the variant with *e marks an action, and the variant with *a marks the reversal, or undoing of that action. Examples include *-bej 'to tie by winding around (as in binding up a package)', but *-baj 'to untie, unravel, unwind', *-keb 'lid, cover; to cover', but *-kab 'to open, uncover', and *-kes 'to encircle, wrap firmly around (as a snake)', and *-kas-$_2$ 'to loosen, undo, untie'. This is particularly intriguing, since it appears to operate only on the level of the root, there being no known morphological evidence for an *e/a contrast with this functional distinction.

2.7 Referents of roots

It is an open question why some roots are so much more robustly attested than others. Examples such as *-Cuk 'to knock, pound, beat, sound of these actions', *-kaŋ 'to spread apart, as the legs', *-keC 'adhesive, sticky', *-pit 'to press, squeeze together; narrow', or the family of roots that begin with *ku- and mean 'bend, curve' were recognized early because of their prominence in the lexicons of many languages. In general, the most robustly attested roots appear to be those that reference sensory impressions about the world of practical (work) activities, or experiences in the natural world (actions or states of bending, spreading in a 'V' shape, sticking or adhering, pressing against, and the like).

As already noted, probably the most obvious division of roots is into those that are onomatopoetic, and those that are not. Although the boundary between these is usually quite clear, a root that symbolizes a natural sound may also appear in morphemes that refer to the action performed, or occasionally to the instrument used to produce that sound. Thus PMP *bak 'clap! smack!', Dairi-Pakpak Batak gəbak-gəbak 'the sound of wings flapping', and Central Tagbanwa rabak 'the sound of rain, tapping, or rapid gunfire' are clearly intended to represent a particular type or quality of sound, while Mandar rimbaʔ 'to beat the wings' or Maranao robak 'to hammer; sledgehammer', which appear to contain the same root, focus more on the action performed or the instrument used to produce this kind of sound. In some cases cultural information may be needed to see the connection between an action and the sound it typically produces. Words for 'to wash clothes', for example, are sometimes formed with *-pak 'to slap, clap' because clothes were traditionally soaked in a river, and then beaten with a wooden mallet on a river rock to help loosen the dirt before the final rinse.

Non-onomatopoetic roots show greater variation in the referents represented. Appendix 1 lists numbers of separate roots for each category of meaning in two types of display. The first of these arranges the semantic categories represented by the roots collected here in three columns: verbs, nouns, and others that are either adjectives or stative verbs, which I will call 'statives'. The number immediately following the generalized gloss indicates the number of distinct roots that are associated with that meaning, and the number in parentheses shows the number of EIMs in which these roots have been found to date. While this provides a quick overview of the breakdown by word class, it has the disadvantage of separating semantic categories that are spread over more than one word class, as with 'to bend, curve', 'a bend, curve (as in a path/road)', and 'bent, curved'. To address this problem the second display arranges roots by number of attestations for a given meaning, from highest to lowest, ignoring word class distinctions. This has the advantage of showing at a glance which types of meanings are most robustly attested among the -CVC roots that have been identified to date in at least four etymologically independent examples.

The following observations can be made based on the material in Appendix 1. First, about 60 of the 406 roots recorded so far, or roughly 15%, are arguably onomatopoetic, while 185 of the 406 non-onomatopoetic roots recorded so far, or around 46%, are dynamic verbs. The remaining 39% of roots are about even divided between nouns and adjectives or stative verbs, hence about 20%

in each category. Second, the most commonly represented semantic categories tend to be aspects of physical perception, with 'bend, curve' the most common (associated with 10 distinct roots and 206 EIMs), followed by 'shine, glitter' and 'sink, submerge' (each associated with 7 distinct roots). These figures represent the relationship between roots and semantic categories or referents. However, the figures in parentheses represent the relationship between roots and morphemes. Some of these are quite surprising, and take us back to the question how to distinguish root-sharing from doubleting. As noted earlier, Blust (2011) treated PMP *ñamuk and *lamuk 'mosquito' as doublets no different than e.g. PMP *tiduR and PAN *tuduR 'to sleep'. However, in pursuing the matter further, 12 EIMs that end with a reflex of *-muk have been found meaning 'mosquito' or 'sandfly', and the similarities connecting this set of words must now accordingly be viewed as due to root-sharing rather than doubleting. This in itself raises the question how large a set of doublets can be. Although the term itself suggests two members, Blust (2011:437) gives one doublet family with eight: *izap, *ki(n)sap, *kizab, *kezem, *kizem, *kizep, *kedip, *kelip 'to blink, wink, flash'. However, since *kezem and *kizem are now considered to contain a root *-zem, found also in PWMP *pezem 'to close the eyes', and Mukah Melanau *pajəm* 'closed, as the eyes; extinguished, of a fire', it appears likely that other members of this set will ultimately be found to share a common root.

Perhaps even more puzzling than differences in the robustness of attestation is the question why recurrent partials exist in complex morphemes for some semantic categories, but not others. Why are well-established reconstructions like PMP *ñamuk, or *lamuk 'mosquito' part of a larger root family that includes at least ten other morphemes of the same meaning that end with *-muk, when e.g. *manuk 'chicken; bird' is a simple morpheme with no known submorphemic connection to any other? Needless to say, this is a question that can be asked about hundreds or even thousands of forms, since the great majority of morphemes in AN languages are simple, not complex.

Finally, it may be worth noting that although several roots express the names of animals, all of these seem to refer to pests, as seen in *-baw 'rat, mouse', *-muk 'mosquito', *-ŋaw-$_1$ 'fly (insect)', and *-ŋaw-$_2$ 'rice bug, insect destructive to crops'. Only one root has been found that refers to plants, namely *-Ceŋ 'stinging nettle: *Laportea* spp.'. This root is also unusual in that four of the five morphemes in which it appears agree in the last -CVCVC, and all five may agree if Amis *lidataŋ* shows metathesis of the first two consonants, and *d* irregularly reflects *d rather than *N, which appears to be the case in Amis *lidoŋ* 'shelter; shadow, shade' (root: *-duŋ 'to shelter from rain or sun; head cover'). Curiously, just as every animal term for which a root has been recorded can be characterized as a pest, the single plant that is represented in forms with a root is a nuisance, since it can cause a painful sting when touched.

2.8 The origin of roots

Given their ubiquity in the modern languages, one must must ask "What is the origin of -CVC roots?". This question was addressed briefly in Blust (1988b), where no firm conclusion could be reached. Lewitz and Jenner (1973) compared the final syllable of Austronesian forms with the base morphemes of Mon-Khmer languages in the hope of finding evidence of recurrent sound correspondences, but without significant results. Similarly, Sagart (1994) assumed that the bound roots of attested AN languages were free forms at some earlier time, and used them as evidence for his controversial 'Sino-Austronesian' hypothesis. However, there is little convincing evidence that non-onomatopoetic roots ever existed as free forms. The only other suggestion offered to date was proposed by J.C. Anceaux, who thought they could be products of blending (Blust 1988b:53–54), but there are also problems with this hypothesis. The answer to this question remains as open today as it was in 1988.

2.9 False cognates?

The term 'false cognate' is usually used for phonetically and semantically similar words that do not show recurrent sound correspondences (and hence have arisen by chance). As such, it can apply either to words in languages for which a relationship has not been established, as with the well-known case of German *nass*, Zuni *nas* 'wet' (Ruhlen 1987:11), or in related languages, where they sometimes present greater difficulty.

Brandstetter (1916) took pains to explain the major sound correspondences between the languages he compared to ensure his readers that the material he was comparing was related, and not a product of chance. I have done less of that here, both because I am using material from many more languages, but also because the sound correspondences can easily be seen at a glance by inspecting comparisons in Blust and Trussel (2020), which is freely accessible online. However, a few words in relation to 'lookalikes' are perhaps in order before addressing the theoretically more challenging problem of distinguishing valid cognate sets from spurious comparisons that exhibit recurrent sound correspondences in complex morphemes.

At first glance Uma -*huŋkaʔ*, Rotinese *huka* 'to open' may appear to be part of a single cognate set, and hence should not be counted separately as evidence for the root *-kaq 'to open forcibly'. However, the /h/ in these languages has different sources (*s in Uma and *p in Rotinese). Much the same can be said for other examples that share a phonetic and semantic similarity by chance, as Kavalan *tabuk* 'dust as gathered in a house', Arosi *tahu* 'bore holes; wood-boring insect' (where Kavalan /t/ must reflect *t, and Arosi /t/ reflects *s), or PWMP *beluŋ 'bend, curve', BM *beluŋ* 'dented, bent, curved', where PWMP *e is a schwa, which almost invariably became BM /o/. These are false cognates in the traditional sense of the term.

What the presence of roots in AN morphemes introduces is a more insidious definition of 'false cognate', namely one in which morphemes of the same or similar meaning show recurrent sound correspondences that have arisen by convergence, and so are *a priori* indistinguishable from true cognates that have arisen through divergence from a common ancestor. What is arguably the most challenging problem associated with recognition of the root in AN languages is not concerned with the root itself, then, but rather with how complex morphemes (i.e. those that contain a recurrent submorphemic sound-meaning association) undermine our confidence in the comparative method.

The comparative method is designed to reconstruct phonology, and phonological reconstruction is only possible through the reconstruction of morphemes. In classical accounts, as that of Saussure (1915) the morpheme is an arbitrary association of sound and meaning with no internal meaningful parts. It is therefore the smallest unit of meaning in language. Over time morphemes may change their phonemic shapes and meanings as the communities descended from a common ancestor separate, and the languages they speak diverge. A cognate set, then, is a collection of morphemes that have arisen through a process of divergence, and by application of the comparative method a proto-form, the ancestral type of each cognate set, can be reconstructed or inferred with a high degree of confidence.

What gives us this confidence in the universal applicability of the comparative method? Put simply, there are two possible explanations for recurrence in the sound-meaning associations of lexical items in related languages, one of which can be called the 'convergence hypothesis' and the other the 'divergence hypothesis'. Given a set of forms of related meaning like Puyuma *laŋit*, Ilokano *láŋit*, Kayan *laŋit*, Malay *laŋit*, Tae' *laŋiʔ*, Chamorro *laŋet*, Palauan *yaŋd*, Yamdena *laŋit*, Vitu *laŋi*, Wogeo *laŋ*, Chuukese *nááŋ*, Samoan *laŋi*, Maori *raŋi*, Hawaiian *lani*, all meaning 'sky, heaven', one sees a recurrent similarity, that is paralleled by recurrent sound correspondences across other morphemes (Puyuma *taŋis*, Kayan *taŋih*, Malay *taŋis*, Tae' *taŋiʔ*, Chamorro *taŋes*, Samoan *taŋi*, Hawaiian *kani*, etc. 'to weep, cry'). The first logically possible explanation for this similarity is that these forms have arisen through historically independent innovations in each language—that is, through convergence. However, there is nothing in the referent 'sky, heaven' that requires speakers of different languages to innovate a similar sequence of sounds, to represent this meaning, since if there were we must ask why words of this general shape do not represent the meaning 'sky, heaven' in all of the world's languages, and similarly with countless other examples. The convergence hypothesis therefore fails to account for a basic observation—that some languages exhibit systematic lexical similarities with one another that do not occur at all with other languages.

We are left, then, with the divergence hypothesis. Languages that exhibit systematic similarities are grouped into families, each of which is assumed to have developed from a single ancestral language community, and the cognate sets that exhibit these similarities are assumed to have developed through a process of divergence from a common ancestral morpheme, or proto-form, in this case PAN *laŋiC 'sky, heaven'. What is critical to recognize for present purposes—but is rarely mentioned—is that the divergence model assumes that morphemes cannot contain smaller meaningful parts. This is true of simple morphemes such as PAN *laŋiC, or the vast majority of other morphemes in any human language, which is why the comparative method works in these cases—since the sound-meaning association is largely arbitrary, systematic similarity across multiple morphemes cannot plausibly be attributed to chance, and therefore implies historical contact.

However, we have known for nearly a century that not all morphemes are simple, since submorphemic sound-meaning associations exist in many languages, as seen in English words like *glare, gleam, glimmer, glint, glisten, glitter,* or *glow*, where the sequence *gl-* refers recurrently to light, *sniff, sniffle, sneer, sneeze, snicker, snore, snot, snout* and the like, where *sn-* refers to the nose, or *bash, clash, crash, dash, smash*, where *-sh* signals violent collision. Firth (1930) called these 'phonaesthemes' (written 'phonesthemes' in American English). So long as our focus is entirely on description, complex morphemes present no problem to linguistic theory – they simply require that the phonestheme, which in Indo-European languages consists of initial consonant clusters, or single final consonants, be extended to include complete -CVC syllables. But what happens when complex morphemes are viewed in a comparative context?

In compiling the data for this study, the corpus of complex morphemes has sometimes become 'crowded' with forms that differ etymologically in only a single vowel or consonant. This is particularly true with robustly supported roots such as *-Cuk 'to knock, pound, beat; the sound of these actions', *-kaŋ-$_2$ 'to spread apart, as the legs', or *-pit 'to press, squeeze together; narrow'. In some cases the crowding has become so intense that forms in widely separated languages show sound correspondences that appear to indicate a common proto-form, although the meanings may be quite divergent, raising the question whether these comparisons are products of divergence, as with true cognates, or have developed through convergence because they share a common root. To make this clearer, the examples (1–25) as listed below present some sample comparisons that illustrate how the comparative method, which is fully trustworthy when working with simple morphemes, may not be trustworthy when working with morphemes that contain a shared root.

Lexical comparisons with a 'complex' morpheme: Valid sets or false cognates?

1) *-bek-$_1$ 'dull muffled sound': Ifugaw *imbók* 'beating of gongs while people dance', Tagalog *hibík* 'to sob'. Reconstruction: PPH *hi(m)bék (GLOSS?).
2) *-bek-$_2$ 'rotten, crumbling, pulverized': Itbayaten *axbək* 'dust, dust particles', Agutaynen *lebek* 'to pound rice with mortar and pestle'. Reconstruction: PPH *lebek (GLOSS?).
3) *-bit 'to hook, clasp, grasp with fingers': PAN *kabit 'hook', PPH *kabit 'to lead by the hand, support, as a feeble person. Reconstruction: (combine, or keep separate?).
4) *-Cuk 'to knock, pound, beat': Komodo *weto?* 'woodpecker', Manggarai *wetuk* 'to hit with something'. Reconstruction: Proto-? *(bw)etuk (GLOSS?).
5) *-dem 'dark; overcast': PMP *kidem 'to close the eyes', Ilokano *kiddém* 'to tarnish, become dull', Hawaiian *kilo* 'first night of the new moon'. Reconstruction: PMP *kidem (GLOSS?).
6) *-kab 'to open, uncover': PWMP *iŋkab 'to open, uncover', Tagalog *hikáb* 'a yawn', Kelabit *ikab* 'opening in the longhouse roof for the entrance of light and air'. Reconstruction: PWMP *hi(ŋ)kab (GLOSS 'to open, uncover'?).
7) *-kaŋ 'to spread apart, as the legs': Botolan Sambal *halakáŋ* 'to split and spread apart the end of bamboo and make a hen's nest of it', Balinese *səlaŋkaŋ* 'stand astride of'. Reconstruction: PWMP *sala(ŋ)kaŋ (GLOSS?).
8) *-kas 'to loosen, undo, untie': PAN *bekas 'to spring a trap', PMP *beŋkas 'to untie, undo'. Reconstruction: (combine, or keep separate?).
9) *-luR 'to flow': PAN *iluR 'river channel', PMP *iluR 'spittle, saliva'. Reconstruction: (combine, or keep separate?).
10) *-ŋa 'to gape, open the mouth wide': PAN *paŋa 'fork of a branch; any forked structure, bifucation', POC *paŋa 'to gape, be open'. Reconstruction: (combine, or keep separate?).

11) *-ŋus 'snout': Sundanese *baŋus* 'snout, beak, nose of an animal', PMic *faŋus- 'to blow the nose'. Reconstruction: PMP *baŋus (GLOSS?).
12) *-ŋus 'snout': Malay *hiŋus* 'nasal mucus, snot', Tetun *inus* 'nose, trunk, snout'. Reconstruction: PMP *qiŋus (GLOSS?).
13) *-pak 'to slap, clap': Lun Dayeh *rafak* 'the sound of splashing water', Maranao *rampak* 'to hit by ricochet; get two birds with one stone'. Reconstruction: PWMP *rampak (GLOSS?).
14) *-pak 'to slap, clap': Ilokano *tupák* 'the sound of a soft object falling', Lun Dayeh *tufak* 'a kick'. Reconstruction: PWMP *tupak (GLOSS?).
15) *pet 'plugged, stopped, closed off': Javanese *ḍipət* 'tightly closed, of eyes', Bolaang Mongondow *dimpot* 'to plug or cover up crevices, holes in the roof, etc. Reconstruction: PWMP *di(m)pet (GLOSS?).
16) *-pit 'to press, squeeze together; narrow': Iban *jumpit* 'tight, of clothes', Kambera *júmbitu* 'to pinch'. Reconstruction: PMP *zumpit (GLOSS?).
17) *-riC 'the sound of ripping, etc.': Paiwan *gurits* 'to squeal (pig)', Tontemboan *gorit* 'to saw'. Reconstruction: PAN *guriC (GLOSS?).
18) *-sek 'to insert, stick into a soft surface': PAN *hesek-$_2$ 'to drive in stakes or posts', PPH *hesek-$_1$ 'to plant seeds by dibbling'. Reconstruction: (combine, or keep separate?).
19) *-suR 'satiated, full after eating': Tidong (Malinau) *masug* full (stomach), POC *masuR 'satiated, full after eating'.[13]
20) *-tak 'the sound of cracking, splitting, etc.': PMP *le(n)tak 'to clack the tongue', PWMP *letak 'to split, crack'. Reconstruction: (combine, or keep separate?).
21) *-tas 'to sever, rip apart, cut through; short cut': PAN *betas 'to tear, rip open (as cloth or stitches)', PMP *bentas 'to hack a passage through, blaze a trail'. Reconstruction: (combine, or keep separate?).
22) *-tas 'to sever, rip apart, cut through; short cut': PWMP *butas-$_1$ 'to cut through, sever', PWMP *butas-$_2$ 'to separate people, disperse a gathering'. Reconstruction: (combine, or keep separate?).
23) *-tas 'to sever, rip apart, cut through; short cut': Itawis *maf-fútat* 'to wean', Abaknon *putas* 'to cut with one stroke'. Reconstruction: PPH *putas (GLOSS?).
24) *-tek-$_1$ 'a clicking or light knocking sound': PAN *tektek-$_1$ 'chopping to pieces, cutting up, as meat or vegetables', PAN *tektek-$_2$ 'gecko; chirp of a gecko'. Reconstruction: (combine, or keep separate?).

---

[13] I treat the agreement in 19) as convergent because Tidong (Malinau) is the only non-Oceanic language known to contain a form that could reflect *masuR, and given numerous other forms in Sabahan languages that have *na-asug* it seems most likely that this Tidong form is an independent innovation from *ma-asuR.

25) *-tuk 'top, summit, crown': Casiguran Dumagat ʔóntok 'up, top; end; to go up higher, to ascend (of the sun, an airplane, etc.)', Western Bukidnon Manobo *utuk* 'to carry on top of the head'. Recontruction: PPH *u(n)tuk (GLOSS?).

Other examples of questionable cognates that satisfy all requirements of the comparative method could be cited, but these should be sufficient to make the issue clear. Some of these comparisons may be valid and others not—the problem with the comparison of complex morphemes is knowing when one explanation is to be preferred over the other, and decisions about this may well differ among particular investigators.

In rare instances I have proposed a single reconstruction where intuition tells me the forms assigned to it probably are historically independent products of convergence, as with PMP *du(m)pak 'to strike, collide with', since otherwise Tagalog *lupák* 'to pound rice so as to separate grain from chaff', Javanese *ḍupak* 'to kick with the bottom of the heel', Manggarai *dumpak* 'to run aground, collide with', all of which appear to share the root: *-pak 'to slap, clap') would be counted as three EIMs despite sharing recurrent sound correspondences and a generic similarity of meaning. Needless to say, if I did count the forms from these three languages as independent testimony for *-pak, this would increase the tally of supporting evidence beyond the 52 cases already recognized.

As can be seen, all of these comparisons allow the reconstruction of a proto-form, but problems begin when it comes to glossing it, whether we are combining forms in attested languages, or previously proposed homophonous reconstructions. It is true that with sufficient imagination glosses can be provided in some cases, as with 6). However, while there is a generic semantic similarity that is conferred by the shared root, the details differ so much in some cases that one can easily imagine that these phonologically compatible forms are products of parallel histories. This is not to say that we can be sure about this one way or the other, since true cognates sometimes show considerable semantic latitude. The problem at bottom is one of uncertainty—the legitimacy of such comparisons as products of divergence from a common ancestral form is brought into question so sharply that scholars may well be divided on whether to recognize phonologically corresponding forms of similar meaning as cognate or not. Even where a plausible gloss is proposed, as with PWMP *hi(ŋ)kab 'to open, uncover', it can be hard to distinguish it from that of other morphemes that contain the same root, and one is left with the uncomfortable feeling that a superficially legitimate comparison that meets all the basic requirements of the tried and tested comparative method, may actually be a product of parallel evolution.

Still another problem is seen in comparing what appear to be partial vs. full reduplications of the same root over a wide range of languages, as with PAN *babaw 'upper surface', and PMP *bawbaw 'upper surface, top; above'. Here we have clearly related forms built from the root *-baw 'high; upper surface', but the question remains, are reflexes assigned to *babaw, which is attested in 44 languages in the ACD reductions of *bawbaw (attested in 5), or are they historically independent formations that are built upon the root *-baw in the same way that PMP *umbaw 'top part; high', Thao *kafaw* 'high; at a height', Ilokano *rimbáw* 'top, peak, summit', or any other complex morpheme that contains this root is built? If *babaw is treated as a reduction of *bawbaw (despite the difference of proto-language, which is potentially irrelevant) the two must be treated as a single cognate set, and hence as one witness for the root *-baw. However, if they are treated as historically independent formations each is an EIM supporting *-baw. Essentially the same problem arises in connection with PMP *bubuŋ(-an) 'ridge of the roof', but Uma *wumu* 'ridge of the roof (apparently reflecting *buŋbuŋ, not *bubuŋ). In the ACD, and again here, I have assumed that such forms are products of convergence around a shared root, but this is certainly debatable.

In other cases we know with certainty that striking similarities are due to convergence, as with Banggai *kepaŋ* 'paralyzed, lame', Makassarese *keppaŋ* 'crippled, lame', or Madurese *poŋkas* 'beginning', Bare'e *poŋka* 'beginning, introduction', where the forms and meanings agree, but indicate mid-front and mid-back vowels that did not exist in the proto-language for either of these pairs of languages. The only choice left, then, is to treat them as results of convergence favored by a shared root *-paŋ in the first example, and by *-kas in the second. Additionally, some languages have what appear to be non-borrowed doublets, which suggest that they were independently innovated with the same root and a convergent CV- or CVC- formative, as with Cebuano *íkit* 'close together, usually people', *iŋkít* 'fingers which are joined congenitally together' (root: *-kit 'to join along the length'), Maranao *sapəŋ* 'to cover', *sampəŋ* 'a cover, lid' (root: *-peŋ 'to plug up, dam; cover'), Iban *sikaŋ-sikaŋ* 'with legs apart', *siŋkaŋ* 'a step, pace' (root *-kaŋ 'to spread apart, as the legs'), Malay *dəpus* 'the sound of a rush of air through a narrow opening, e.g. from bellows', *dəmpus* 'to puff in one's sleep' (root: *-pus 'the sound of escaping air'), Manggarai *gipes* 'shrunken from crushing pressure', *gimpes* 'sunken (cheeks), shrunken' (root: *-pes 'empty, deflated'), Manggarai *ŋapok* 'the sound of a slap', *ŋapuk* 'the sound of chewing' (root: *-puk 'to thud, snap, crack, break'), or Tetun *doros* 'slippery', *dorus* 'to slide' (root: *-rus 'to slip or slide off').

Data like this shows that appearances may be deceiving: forms that initially appear to be part of a single cognate set may in fact be products of convergent innovation, and so count as independent evidence for the existence of a root. At first sight nearly all of the comparisons cited above suggest that skepticism about cognation among forms that show recurrent sound correspondences is justified only where the meanings of phonologically compatible forms diverge widely. However, even if the meaning is identical, cognation may be questionable when a large contribution to the phonological correspondence of the forms is due to a shared root. For example, although PMP *ñamuk and *lamuk 'mosquito' have been reconstructed on the basis of widespread cognate sets (represented by 84 languages reaching from Ilokano to Maori for the former and 9 languages reaching from Ilokano to Rotuman for the latter in Blust and Trussel 2020), the comparison Kapampangan *amúk*, Ngadha *emu* 'mosquito' points to a third variant, PMP *emuk. Here the meanings are identical, and yet, given the presence of a root *-muk in eleven other EIMs, one must ask whether PMP *emuk actually existed, or whether it is a product of convergent innovation. In this case the only independent innovation needed to mimic a valid cognate set pointing to PMP *emuk is the initial vowel, since the rest of this word is a root that has now been identified in twelve EIMs.

How many other complex morphemes that appear to be cognate are actually products of convergent innovation? This is a difficult question to answer. One indication that has already been suggested as showing that a comparison may represent false cognates rather than valid ones is the degree of semantic divergence: where the glosses are identical or nearly so there is generally less uncertainty about true cognation. However, as just shown, this is not a rule that invariably holds, since valid cognates are known to vary widely in meaning, and some comparisons that agree in both form and meaning but have a limited distribution, as the apparent reflexes of PMP *emuk 'mosquito', may well be products of convergence.

To summarize, valid cognates are due to a historical process of divergence from a common ancestral form, and this can confidently be inferred due to the improbability that recurrent sound correspondences could develop through historically independent innovations of five or six segments in connection with a given meaning. False cognates, on the other hand, could easily arise through historically independent innovations in complex morphemes that differ only in a single segment or two that are not part of a recurrent submorphemic sound-meaning association. While simple morphemes of the same or similar meaning that show recurrent sound correspondences *can* be products of convergence, this is highly unlikely, whereas for complex morphemes the role of chance in producing false cognates increases dramatically. However, since it is difficult to distinguish recurrent sound correspondences produced by convergence from those produced by divergence, I have generally treated cases like Kapampangan *amúk*, Ngadha *emu* 'mosquito', where the meanings are identical or nearly so, as cognate, but where the meanings are a less perfect match, I treat each form as independent evidence for the root in question despite the superficial appearance of cognation. Since linguistics provides fewer controls on the measurement of semantic similarity than of phonological correspondence, it has not been possible to be completely consistent in treating such examples. As can be seen from the reconstructed forms, some decisions have already been made in the ACD about homophonous morphemes that share a root, and they will be followed here. Some readers may object that these decisions are arbitrary, and artificially inflate the perceived support for a given root, but close attention to the data shows that such cases are rare, and have little effect on the overall conclusions reached.

2.10 The problem of transmission

One aspect of AN roots that has been recognized only in Blust (1988b), is how sound-meaning associations below the level of the morpheme can be learned, and thereby transmitted between the generations. When they are embedded in cognate morphemes, as with PMP *labuq 'to fall' (root: *-buq), or *telen 'to swallow' (root: *-len), it is clear that the morphemes in which they appear are learned as part of the acquisition of the lexicon of a language. However, for roots that are known only in morpheme isolates this is a serious problem, since there is no obvious way that they could be acquired by direct transmission except in shared morphemes, which would form part of a cognate set.

Two potential solutions to this problem suggest themselves. The first is that all roots will ultimately be found to occur only in reconstructed forms. However, given the number of roots that are currently known only in morpheme isolates, and the likelihood that many reconstructed forms with monosyllabic roots are based on false cognation, this possibility does not appear very promising. The second solution is that in addition to learning the lexicon of a language speakers acquire a largely subliminal awareness of phonesthemes that can then be extended to new cases. In a language like English this would mean that new coinages relating to light in its various forms would be more likely than not to begin with *gl-*, new coinages relating to the nose would be more likely than not to begin with *sn-*, and so forth because speakers have internalized an unconscious sound-meaning association based on its recurrence across a word family.

However, this still leaves many questions unanswered. First, how can a 'subliminal awareness' (as opposed to an actual linguistic form) be transmitted to children learning a language? Is it simply a by-product of learning vocabulary that can then be used to form neologisms? Second, how do phonesthemes like *gl-*, or *sn-* start in the first place? Are they triggered by the presence of just two psychologically prominent morphemes that happen by chance to share a sound-meaning association that is then generalized to new lexical innovations? If this is the case for AN roots, one must still ask why they tend to be overwhelmingly final -CVC sequences, since accidental sound-meaning associations in two morphemes that might initiate such a cascade of changes could occur anywhere in a CVCVC or CVCCVC morpheme. Add to this puzzle that fact that some roots (almost always the least robustly attested ones) are found only in morpheme isolates, and one must ask whether it is even necessary for a language to inherit one or more complex morphemes in order to develop new complex morphemes that contain a given root. The root *-baw 'shallow', for example, is represented by a single reconstructed form, PWMP *babaw 'shallow', but is attested to date in twelve other morphemes. How was it extended to these other supporting examples if the model for it was a single form? As noted above, still other roots are not yet known in any proto-form, as *-deŋ 'straight; to straighten', which is attested in the data collected here in seven morpheme isolates across a range of languages reaching from northern Luzon to northern Sumatra. How could this submorphemic sound-meaning association possibly be transmitted across many generations of speakers with no inherited morpheme to carry it? Given these and other still unanswered questions it seems clear that the problem of transmission is a major conundrum that is yet to be adequately understood.[14]

---

[14] Bergen (2004) cites experimental evidence which suggests that English phonesthemes are recognized and processed no differently than morphemes, but this claim still leaves many questions unanswered in relation to the Austronesian data.

3 Summary of the data

For ready reference, a complete list of the 406 monosyllabic roots identified in the present work is given below in two displays: first in alphabetical order, and second by order of magnitude based on supporting examples. For various reasons, attempts to present a summary in double columns were unsuccessful, and—despite its greater length—having a separate display for order of magnitude also makes it easier to compare the current material with that in Blust (1988b), which was presented in a similar fashion. In all, the 406 roots proposed here are supported by 5,401 EIMs, for an average of about 13 supporting examples per root. In Display 2, OC = order class and EIMs = number of etymologically independent morphemes in which a root has been found. The full data with supporting evidence appears in Chapter 4.

Display 1: Roots in alphabetical order.

1.	*-bag	the sound of a heavy smack (10)
2.	*-baj	to untie, unravel, unwind (13)
3.	*-bak$_1$	peeling skin (12)
4.	*-bak$_2$	the sound of a heavy smack (31)
5.	*-bak$_3$	to split off (11)
6.	*-bak$_4$	valley, ravine, gully (4)
7.	*-ban	group, company (5)
8.	*-baŋ$_1$	broad, wide (11)
9.	*-baŋ$_2$	door, gate (7)
10.	*-baŋ$_3$	to fly (7)
11.	*-baR	wide (7)
12.	*-baw$_1$	high; upper surface (26)
13.	*-baw$_2$	rat, mouse (8)
14.	*-baw$_3$	shallow (13)
15.	*-bay	to accompany; partner (10)
16.	*-beg	thump, dull sound of collision (5)
17.	*-bej	to wind around repeatedly (21)
18.	*-bek$_1$	dull, muffled sound (23)
19.	*-bek$_2$	rotten, crumbling, pulverized (9)
20.	*-beŋ$_1$	to block, stop, dam (13)
21.	*-beŋ$_2$	dull, resounding sound (9)
22.	*-bet	buttocks (5)
23.	*-bid	to twist, twine together (14)
24.	*-bir	rim, edge (11)
25.	*-bis	to drip (6)
26.	*-bit$_1$	to hook, clasp; grasp with fingers (33)
27.	*-bit$_2$	a whip; to whip (6)
28.	*-buC	to weed, pluck, pull out (26)

Display 1 (continued)

29.	*-bud	to sow, scatter seed in planting (4)
30.	*-buk-$_1$	to decay, crumble; powder (28)
31.	*-buk-$_2$	to pound; a thud, heavy splash (34)
32.	*-bun-$_1$	drizzle, mist, fog; hazy vision (5)
33.	*-bun-$_2$	to heap up, cover with earth; collect, gather (29)
34.	*-buŋ-$_1$	a deep resounding sound (11)
35.	*-buŋ-$_2$	proud, haughty (4)
36.	*-buŋ-$_3$	roof ridge (11)
37.	*-buq-$_1$	to grow, increase (9)
38.	*-buq-$_2$	to fall (7)
39.	*-bur	to stir, mix (6)
40.	*-buR-$_1$	to mix together, stir (12)
41.	*-buR-$_2$	rice gruel; rice porridge (9)
42.	*-buR-$_3$	to strew, sow; sprinkle (13)
43.	*-buR-$_4$	turbid (12)
44.	*-bus.$_1$	to end; finished, used up (11)
45.	*-bus.$_2$	to leak, spill out, as through a hole (11)
46.	*-but.$_1$	buttocks, bottom (7)
47.	*-but-$_2$	hole (8)
48.	*-but.$_3$	husk, coarse hair or fiber (6)
49.	*-but-$_4$	unclear, hazy, misty (4)
50.	*-bux	dust (15)
51.	*-cak-$_1$	muddy; the sound of walking in sticky mud (19)
52.	*-cak-$_2$	to stab (5)
53.	*-cek	blind (5)
54.	*-ceq	to hatch, break into large pieces (10)
55.	*-cik	to fly out, splash, spatter (37)
56.	*-cit	to squirt out (25)
57.	*-cut	to squirt, squeeze or slip out (20)
58.	*-Caq	mud; earth, ground (8)
59.	*-Ceg	to hit, beat, pound (9)
60.	*-Cek	mottled pattern (9)
61.	*-Cem	dark, of color; a dark color (10)
62.	*-Ceŋ	stinging nettle: *Laportea* spp. (5)
63.	*-Cik-$_1$	mottled, spotted (18)
64.	*-Cik-$_2$	to spring up; flicking motion (27)
65.	*-Cug	to knock, pound, beat (15)
66.	*-Cuk	to knock, pound, beat; sound of these actions (58)
67.	*-Cuŋ	deep resounding sound (19)
68.	*-daŋ-$_1$	to dazzle, shine (8)
69.	*-daŋ-$_2$	to warm by a fire (9)
70.	*-daR	to lean on, recline (9)
71.	*-dek-$_1$	to hiccough, sob (10)
72.	*-dek-$_2$	pulverized, pounded fine (7)
73.	*-dem-$_1$	dark; overcast (40)
74.	*-dem-$_2$	to brood; begrudge; remember; keep still (18)

Display 1 (continued)

75. *-deŋ	straight; to straighten (7)	
76. *-det	packed in, compressed (13)	
77. *-dik	small (7)	
78. *-diR	to lean against (4)	
79. *-du	ladle, spoon (4)	
80. *-duk	ladle, spoon (11)	
81. *-duŋ	to shelter from rain or sun; head cover (38)	
82. *-duR	thunder; the rumbling of thunder (6)	
83. *-gak-$_1$	proud; to boast (10)	
84. *-gak-$_2$	raucous, throaty sound (19)	
85. *-gaŋ	dry (13)	
86. *-gaw	confused, disoriented, lost (6)	
87. *-gek	dull throaty sound (17)	
88. *-gem	to grasp in the fist (16)	
89. *-geŋ	to hum, buzz (9)	
90. *-ger	to shake, shiver, tremble (6)	
91. *-get	angry (8)	
92. *-gik	a shrill throaty sound (10)	
93. *-gis	to scrape (4)	
94. *-git	anger, resentment (5)	
95. *-guC	to pull with a jerk (11)	
96. *-guk	a deep throaty sound (25)	
97. *-guŋ	a deep resounding sound (21)	
98. *-gur	to purr, rumble (16)	
99. *-gut-$_1$	angry, annoyed (4)	
100. *-gut-$_2$	to gnaw (9)	
101. *-hak	clearing of throat (4)	
102. *-huR	to mix, as food with water (5)	
103. *-ka	to split, force open (8)	
104. *-kab-$_1$	to open, uncover (18)	
105. *-kab-$_2$	to snap at with the teeth (6)	
106. *-kad	to prop, support (6)	
107. *-kak	to cackle, laugh loudly (26)	
108. *-kaŋ-$_1$	to bark, croak (6)	
109. *-kaŋ-$_2$	to spread apart, as the legs (58)	
110. *-kaŋ-$_3$	stiff, rigid; cramps (8)	
111. *-kap	to grope, feel in the dark, etc. (17)	
112. *-kaq-$_1$	to open forcibly (18)	
113. *-kaq-$_2$	to split (21)	
114. *-kas-$_1$	to begin (9)	
115. *-kas-$_2$	to loosen, undo, untie (43)	
116. *-kas-$_3$	quick, agile, strong, energetic (21)	
117. *-kat	to rise, climb (14)	
118. *-kaw-$_1$	a curve; to curve; circuitous (5)	
119. *-kaw-$_2$	high, tall (9)	
120. *-keb-$_1$	a lid, cover; to cover (19)	

Display 1 (continued)

121.	*-keb-$_2$	to lie face down, be prone (15)
122.	*-keC	adhesive, sticky (56)
123.	*-ked	to prop, support; staff (11)
124.	*-kek	to shriek, creak, cluck, chuckle (12)
125.	*-kel-$_1$	to bend, curl (15)
126.	*-kel-$_2$	harsh coughing (10)
127.	*-kem-$_1$	to clench, cover; grasp (25)
128.	*-kem-$_2$	to lie face down, be prone (5)
129.	*-keŋ-$_1$	cramps, stiffening of limbs (10)
130.	*-keŋ-$_2$	hollow, resounding sound (6)
131.	*-keŋ-$_3$	to shrink, shrivel (5)
132.	*-kep-$_1$	to cover; fold over (23)
133.	*-kep-$_2$	to seize, grasp, embrace (25)
134.	*-keR	stiff, rigid, as a corpse (6)
135.	*-kes	to encircle, wrap firmly around (28)
136.	*-kik	a shrill throaty sound (24)
137.	*-kiŋ	a clear ringing sound (11)
138.	*-kiq	a high-pitched vocal sound (8)
139.	*-kis-$_1$	to scratch, grate, scrape (8)
140.	*-kis-$_2$	to tie, bind, wrap around (8)
141.	*-kit-$_1$	to bite (11)
142.	*-kit-$_2$	to join along the length (27)
143.	*-kit-$_3$	to remove, detach (5)
144.	*-kit-$_4$	small (6)
145.	*-kub.$_1$	a cover; to cover (22)
146.	*-kub.$_2$	to surround, encircle (12)
147.	*-kuC	hunched over, bent (26)
148.	*-kud	cane, staff, walking stick (12)
149.	*-kug	curl, curve (16)
150.	*-kuk-$_1$	bent, crooked (31)
151.	*-kuk-$_2$	sound of a sob, cackle, etc. (21)
152.	*-kul-$_1$	to curl, bend (24)
153.	*-kul-$_2$	snail (4)
154.	*-kum	to close by folding (6)
155.	*-kuŋ-$_1$	to bend, curve (48)
156.	*-kuŋ-$_2$	deep resounding sound (18)
157.	*-kup	to enclose, cover (32)
158.	*-kuq	bend, curve (32)
159.	*-kur	to coo; turtledove (13)
160.	*-kus	to wind around; bundle (13)
161.	*-kux	bend, curve (23)
162.	*-laC	to shine; flickering or flashing light (13)
163.	*-laj	to spread out to dry in the sun (7)
164.	*-lak-$_1$	to shine (16)
165.	*-lak-$_2$	to split (9)
166.	*-laŋ-$_1$	to glitter, flash (11)

Display 1 (continued)

167.	*-laŋ-$_2$	intervening space (8)
168.	*-laŋ-$_3$	spotted, striped (9)
169.	*-lap	to flash, sparkle (35)
170.	*-laq-$_1$	to split (16)
171.	*-laq-$_2$	tongue; to lick (8)
172.	*-lat-$_1$	to open the eyes wide (15)
173.	*-lat-$_2$	scar (11)
174.	*-law	dazzling light (13)
175.	*-leb	to sink, disappear under water (12)
176.	*-leC	interval, gap, intervening space (8)
177.	*-led	to sink (6)
178.	*-lem-$_1$	dark; obscure (42)
179.	*-lem-$_2$	in, inside; deep (10)
180.	*-lem-$_3$	to sink, submerge (8)
181.	*-len	to swallow (6)
182.	*-leŋ	sound that stuns a person (6)
183.	*-lep-$_1$	shiny, bright (6)
184.	*-lep-$_2$	to sink, submerge (7)
185.	*-leR	defective vision (4)
186.	*-liŋ-$_1$	clear ringing sound (19)
187.	*-liŋ-$_2$	cross-eyed (7)
188.	*-liŋ-$_3$	to tilt to the side (6)
189.	*-liŋ-$_4$	to turn, revolve, turn around (18)
190.	*-lip	to insert, slip in (4)
191.	*-liq	to return, restore (6)
192.	*-liR	to flow (7)
193.	*-lit-$_1$	to caulk; adhesive material (18)
194.	*-lit-$_2$	difficult, complicated (4)
195.	*-lit-$_3$	to wind, twist (12)
196.	*-luk	bend, curve (28)
197.	*-luŋ-$_1$	bend, curve (20)
198.	*-luŋ-$_2$	to shelter, shade (20)
199.	*-luR	to flow (12)
200.	*-lus	to slip off, slide down (25)
201.	*-lut	to twist around (5)
202.	*-mak	mat (5)
203.	*-mek	to crush, pulverize; powder (16)
204.	*-mes	to squeeze, knead (7)
205.	*-mis	sweet (8)
206.	*-mit	small, slight (5)
207.	*-muk-$_1$	to crush, pulverize; powder (6)
208.	*-muk-$_2$	mosquito, sandfly (12)
209.	*-mul	to suck on s.t. (7)
210.	*-muR-$_1$	dew (5)
211.	*-muR-$_2$	to gargle, rinse the mouth (11)
212.	*-nab	to flood, inundate (5)

Display 1 (continued)

213.	*-naŋ	to shine, sparkle (9)
214.	*-naw	an enclosed body of water (10)
215.	*-nek	sound, of sleep (6)
216.	*-nem	cloud, cloudy (4)
217.	*-neŋ	still, tranquil, calm, as water (11)
218.	*-neR	to hear (11)
219.	*-ni	to hide, conceal oneself or s.t. (4)
220.	*-niŋ	clear, limpid, of water (6)
221.	*-nit	to remove, extract (10)
222.	*-nut	husk, fiber (10)
223.	*-ñam	savory, tasty (9)
224.	*-ñat	to stretch (18)
225.	*-ñaw-$_1$	to melt, liquefy (9)
226.	*-ñaw-$_2$	to wash, bathe, rinse (22)
227.	*-ñeb	to dive; sink, disappear under water (13)
228.	*-ñej	to submerge, drown (13)
229.	*-ñep	to dive; sink, disappear under water (11)
230.	*-ñet	stretchy, elastic (6)
231.	*-ñit	stretchy, elastic (5)
232.	*-ñut	stretchy, elastic; springy (9)
233.	*-NaR	ray of light (21)
234.	*-Naw	clear, pure, of water (5)
235.	*-Neb	door, doorway (8)
236.	*-Neŋ	to stare, look fixedly (15)
237.	*-NiC	to skin, peel off (8)
238.	*-NuC	to pull out, uproot (16)
239.	*-Nuŋ	to stare, look fixedly (5)
240.	*-ŋa	to gape, open the mouth wide (28)
241.	*-ŋab	to gape; open, of the mouth (8)
242.	*-ŋaC	anger, irritation (4)
243.	*-ŋag	to look upward (10)
244.	*-ŋak	to screech, howl (21)
245.	*-ŋaŋ	amazed; gaping (16)
246.	*-ŋap	to open, of the mouth; gaping (19)
247.	*-ŋaq	to gape, open the mouth wide (22)
248.	*-ŋar	to howl, shout, scream (5)
249.	*-ŋaw-$_1$	confused, disoriented, lost (9)
250.	*-ŋaw-$_2$	fly, generic for housefly and others (6)
251.	*-ŋaw-$_3$	leaking air, vapor (5)
252.	*-ŋaw-$_4$	rice bug, insect destructive to crops (9)
253.	*-ŋeC	angry; to gnash the teeth (35)
254.	*-ŋek	to grunt, groan (5)
255.	*-ŋel	deaf (9)
256.	*-ŋeŋ	to buzz, hum (8)
257.	*-ŋeR	to hear; noise (13)
258.	*-ŋet	sweat, perspiration (4)

Display 1 (continued)

259. *-ŋiC	anger, irritation (27)	
260. *-ŋik	a shrill throaty sound (15)	
261. *-ŋiŋ	a shrill buzz or hum; to ring (10)	
262. *-ŋis-$_1$	to bare the teeth (15)	
263. *-ŋis-$_2$	cruel (5)	
264. *-ŋit	to cry out, shriek (5)	
265. *-ŋuC	to mumble, murmur, mutter (23)	
266. *-ŋuk	a deep throaty sound (9)	
267. *-ŋuŋ	deep buzzing or humming (7)	
268. *-ŋur	a low-pitched sound (7)	
269. *-ŋus	snout (20)	
270. *-ŋuy	to swim (6)	
271. *-paD	flat, level (5)	
272. *-pag	to strike, beat (15)	
273. *-pak-$_1$	to break, crack, split (32)	
274. *-pak-$_2$	to peel bark from a tree (6)	
275. *-pak-$_3$	to slap, clap (52)	
276. *-paŋ	lame (6)	
277. *-pap	flattened (8)	
278. *-paq	to spit; thing chewed but not swallowed (11)	
279. *-pas	to tear or rip off (17)	
280. *-paw	exceed, surpass (6)	
281. *-pay-$_1$	to drape over, hang down (6)	
282. *-pay-$_2$	to wave, flap (6)	
283. *-ped	to block, obstruct; constipated (5)	
284. *-peg	to hit, beat (6)	
285. *-pek	to decay, crumble; sound of breaking (23)	
286. *-pel	plug, stopper; to cram (8)	
287. *-pen	tooth (6)	
288. *-peŋ	to plug up, dam; cover (11)	
289. *-pes	empty, deflated (10)	
290. *-pet-$_1$	firefly (7)	
291. *-pet-$_2$	to hold, grasp, catch (5)	
292. *-pet-$_3$	plugged, stopped, closed off (24)	
293. *-pi-$_1$	dream (12)	
294. *-pi-$_2$	to fold (6)	
295. *-pid	to braid, wind together (11)	
296. *-pik	to pat, slap lightly (21)	
297. *-pil	to attach, join (8)	
298. *-pin	protective layer of cloth or clothing (5)	
299. *-piŋ	next to, beside (4)	
300. *-piq-$_1$	to break off a piece (8)	
301. *-piq-$_2$	to fold (13)	
302. *-pis-$_1$	to deflate; empty (12)	
303. *-pis-$_2$	thin, of materials (22)	
304. *-pit-$_1$	near, close to (6)	

Display 1 (continued)

305.	*-pit	to press, squeeze together; narrow (70)
306.	*-puk-$_1$	to thud, snap, crack, break (43)
307.	*-puk-$_2$	dust (21)
308.	*-puk-$_3$	to gather, flock together (14)
309.	*-pul	blunt, dull (7)
310.	*-pun	to assemble, collect, gather (21)
311.	*-puŋ-$_1$	a bunch, cluster (20)
312.	*-puŋ-$_2$	to float (7)
313.	*-puŋ-$_3$	rice flour (4)
314.	*-puq	brittle (5)
315.	*-pur	to mix (4)
316.	*-pus-$_1$	to end, finish; used up (33)
317.	*-pus-$_2$	the sound of escaping air (7)
318.	*-put	to puff, blow hard (11)
319.	*-qit	hook, barb (9)
320.	*-raŋ	bright, of light (5)
321.	*-riC	the sound of ripping, etc. (16)
322.	*-rik	a spot, freckle (12)
323.	*-ris-$_1$	a rustling sound (15)
324.	*-ris-$_2$	to scratch a line (24)
325.	*-ris-$_3$	a slice; to slice (5)
326.	*-rit	to scratch a line (27)
327.	*-rud	to scratch, scrape (14)
328.	*-rus-$_1$	a rustling sound (8)
329.	*-rus-$_2$	to slip or slide off (10)
330.	*-Raŋ-$_2$	clanging sound (7)
331.	*-Raq	red (10)
332.	*-Raw	hoarse (4)
333.	*-ReC	to tighten, as a belt (11)
334.	*-Reŋ	to groan, moan, snore (17)
335.	*-Riŋ	to ring (15)
336.	*-Ris	to drip (8)
337.	*-Rud	to scrape (5)
338.	*-Ruŋ	to roar, rumble (17)
339.	*-sak	ripe, cooked (6)
340.	*-sed	to dive, plunge into water (5)
341.	*-sek-$_1$	to cram, crowd (23)
342.	*-sek-$_2$	to insert, stick into a soft surface (27)
343.	*-sem	sour (6)
344.	*-sep	to sip, suck (11)
345.	*-seq	to wash clothes; wet (11)
346.	*-siŋ	to spin around (4)
347.	*-sir	hissing sound (9)
348.	*-sit	to hiss, sizzle (7)
349.	*-suD	to budge, move a bit (5)
350.	*-suk	to insert, penetrate, enter (37)

Display 1 (continued)

351.	*-suŋ	rice mortar (6)
352.	*-suR	satiated, full after eating (8)
353.	*-tak	mud, earth, ground (12)
354.	*-tak-$_2$	the sound of cracking, splitting, knocking (42)
355.	*-taŋ	clanging sound (9)
356.	*-taq	raw; green, of plants (5)
357.	*-taR	flat, level (5)
358.	*-tas	to sever, rip, cut through; short cut (43)
359.	*-taw-$_1$	to float (15)
360.	*-taw-$_2$	lost, astray (5)
361.	*-tay-$_1$	to hang (7)
362.	*-tay	suspension bridge (14)
363.	*-teb	to prune; graze (8)
364.	*-tek-$_1$	a clicking or light knocking sound (23)
365.	*-tek-$_2$	mud (7)
366.	*-teŋ-$_1$	to hum, drone (7)
367.	*-teŋ-$_2$	to stare, look fixedly (6)
368.	*-teŋ-$_3$	to stretch; taut (12)
369.	*-teq	sap, gummy secretion (12)
370.	*-ter	to shiver, tremble (6)
371.	*-teR	stiff, rigid, as a corpse (4)
372.	*-tes	to tear up (10)
373.	*-tik-$_1$	a gentle curve (6)
374.	*-tik-$_2$	small, little, few (9)
375.	*-tik-$_3$	a ticking sound (52)
376.	*-til-$_1$	to pinch, pinch off a bit (4)
377.	*-til-$_2$	a small protruding part (10)
378.	*-tiŋ	a clear ringing sound (17)
379.	*-tip	tongs; pinch together (5)
380.	*-tiR	to tremble, quiver (4)
381.	*-tuk-$_1$	beak (7)
382.	*-tuk-$_2$	bend, curve (16)
383.	*-tuk-$_3$	to nod (6)
384.	*-tuk-$_4$	the top, summit, crown (9)
385.	*-tul	to bounce (4)
386.	*-tuq	to fall (8)
387.	*-tus	to break under tension (13)
388.	*-tut	flatulence; stench (15)
389.	*-waŋ	wide open space (31)
390.	*-waq	spider (19)
391.	*-wit	a hook; hook shaped (25)
392.	*-yuŋ	to shake, sway, stagger (8)
393.	*-yut	bag, carrying pouch (4)
394.	*-zak	to step, tread, trample (23)
395.	*-zam	to borrow (4)
396.	*-zaŋ	long, tall (8)

Display 1 (continued)

397.	*-zeg	to stand erect (20)
398.	*-zek	step, tread, trample (5)
399.	*-zel	dull, blunt (5)
400.	*-zem	to close the eyes (5)
401.	*-zep	to blink; flicker, of light (5)
402.	*-zeR	to stand erect (4)
403.	*-zul	to thrust out, protrude (5)
404.	*-zur	to thrust out, protrude (12)
405.	*-zut-$_1$	to pinch, pick up with thumb and forefinger (11)
406.	*-zut-$_2$	to uproot, pluck, pull out (11)

Display 2: Roots in order of support.

	OC		EIMs	
1.	1	*-pit-$_1$	70	to press, squeeze together; narrow
2.	2	*-Cuk	58	to knock, pound, beat
3.	2	*-kaŋ	58	to spread apart, as the legs
4.	3	*-keC	56	adhesive, sticky
5.	4	*-pak	52	to slap, clap
6.	4	*-tik	52	a ticking sound
7.	5	*-kuŋ-$_1$	48	to bend, curve
8.	6	*-kas-$_2$	43	to loosen, undo, untie
9.	6	*-puk-$_1$	43	to thud, snap, crack, break
10.	6	*-tas	43	to sever, rip, cut through; short cut
11.	7	*-lem-$_1$	42	dark; obscure
12.	7	*-tak-$_2$	42	the sound of cracking, splitting, knocking
13.	8	*-dem-$_1$	40	dark; overcast
14.	9	*-duŋ	38	to shelter from rain or sun; head cover
15.	10	*-cik	37	to fly out, splash, spatter
16.	10	*-suk	37	to insert, penetrate, enter
17.	11	*-ŋeC	35	angry; to gnash the teeth
18.	11	*-lap	35	to flash, sparkle
19.	12	*-buk-$_2$	34	to pound; a thud, heavy splash
20.	13	*-bit-$_1$	33	to hook, clasp; grasp with fingers
21.	13	*-pus-$_1$	33	to end, finish; used up
22.	14	*-kup	32	to enclose, cover
23.	14	*-kuq	32	bend, curve
24.	14	*-pak-$_1$	32	to break, crack, split
25.	15	*-bak-$_2$	31	the sound of a heavy smack
26.	15	*-kuk-$_1$	31	bent, crooked
27.	15	*-waŋ	31	wide open space
28.	16	*-bun-$_2$	29	to heap up, cover with earth; collect, gather
29.	17	*-buk-$_1$	28	to decay, crumble; powder

Display 2 (continued)

	OC		EIMs	
30.	17	*-kes	28	to encircle, wrap firmly around
31.	17	*-luk	28	bend, curve
32.	17	*-ŋa	28	to gape, open the mouth wide
33.	18	*-Cik-$_2$	27	to spring up; flicking motion
34.	18	*-kit-$_2$	27	to join along the length
35.	18	*-ŋiC	27	anger, irritation
36.	18	*-rit	27	to scratch a line
37.	18	*-sek-$_2$	27	to insert, stick into a soft surface
38.	19	*-baw.$_1$	26	high; upper surface
39.	19	*-buC	26	to weed, pluck, pull out
40.	19	*-kak	26	to cackle, laugh loudly
41.	19	*-kuC	26	hunched over, bent
42.	20	*-cit	25	to squirt out
43.	20	*-guk	25	a deep throaty sound
44.	20	*-kem-$_1$	25	to clench, cover; grasp
45.	20	*-kep-$_2$	25	to seize, grasp, embrace
46.	20	*-lus	25	to slip off, slide down
47.	20	*-wit	25	a hook; hook shaped
48.	21	*-kik	24	a shrill throaty sound
49.	22	*-kul-$_1$	24	to curl, bend
50.	21	*-pet-$_3$	24	plugged, stopped, closed off
51.	21	*-ris-$_2$	24	to scratch a line
52.	22	*-bek.$_1$	23	dull, muffled sound
53.	22	*-kep-$_1$	23	to cover; fold over
54.	22	*-kux	23	bend, curve
55.	22	*-ŋuC	23	to mumble, murmur, mutter
56.	22	*-pek	23	to decay, crumble; sound of breaking
57.	22	*-sek-$_1$	23	to cram, crowd
58.	22	*-tek-$_1$	23	a clicking or light knocking sound
59.	22	*-zak	23	to step, tread, trample
60.	23	*-kub.$_1$	22	a cover; to cover
61.	23	*-ñaw-$_2$	22	to wash, bathe, rinse
62.	23	*-ŋaq	22	to gape, open the mouth wide
63.	23	*-pis-$_2$	22	thin, of materials
64.	24	*-bej	21	to wind around repeatedly
65.	24	*-guŋ	21	a deep resounding sound
66.	24	*-kaq-$_2$	21	to split
67.	24	*-kas-$_3$	21	quick, agile, strong, energetic
68.	24	*-kuk-$_2$	21	sound of a sob, cackle, etc.
69.	24	*-NaR	21	ray of light
70.	24	*-ŋak	21	to screech, howl
71.	24	*-pik	21	to pat, slap lightly
72.	24	*-puk-$_2$	21	dust
73.	24	*-pun	21	to assemble, collect, gather

Display 2 (continued)

	OC		EIMs	
74.	25	*-cut	20	to squirt, squeeze or slip out
75.	25	*-luŋ-₁	20	bend, curve
76.	25	*-luŋ-₂	20	to shelter, shade
77.	25	*-ŋus	20	snout
78.	25	*-puŋ-₁	20	a bunch, cluster
79.	25	*-zeg	20	to stand erect
80.	26	*-cak-₁	19	muddy; the sound of walking in sticky mud
81.	26	*-Cuŋ	19	deep resounding sound
82.	26	*-gak-₂	19	raucous, throaty sound
83.	26	*-keb-₁	19	a lid, cover; to cover
84.	26	*-liŋ-₁	19	clear ringing sound
85.	26	*-ŋap	19	to open, of the mouth; gaping
86.	26	*-waq	19	spider
87.	27	*-Cik-₁	18	mottled, spotted
88.	27	*-dem-₂	18	to brood; begrudge; remember; keep still
89.	27	*-kab-₁	18	to open, uncover
90.	27	*-kaq-₁	18	to open forcibly
91.	27	*-kuŋ-₂	18	deep resounding sound
92.	27	*-liŋ-₄	18	to turn, revolve, turn around
93.	27	*-lit-₁	18	to caulk; adhesive material
94.	27	*-ñat	18	to stretch
95.	28	*-gek	17	dull throaty sound
96.	28	*-kap	17	to grope, feel in the dark, etc.
97.	28	*-pas	17	to tear or rip off
98.	28	*-Reŋ	17	to groan, moan, snore
99.	28	*-Ruŋ	17	to roar, rumble
100.	28	*-tiŋ	17	a clear ringing sound
101.	29	*-gem	16	to grasp in the fist
102.	29	*-gur	16	to purr, rumble
103.	29	*-kug	16	curl, curve
104.	29	*-lak	16	to shine
105.	29	*-laq-₁	16	to split
106.	29	*-mek	16	to crush, pulverize; powder
107.	29	*-NuC	16	to pull out, uproot
108.	29	*-ŋaŋ	16	amazed; gaping
109.	29	*-riC	16	the sound of ripping, etc.
110.	29	*-tuk-₂	16	bend, curve
111.	30	*-bux	15	dust
112.	30	*-Cug	15	to knock, pound, beat
113.	30	*-keb-₂	15	to lie face down, be prone
114.	30	*-kel-₁	15	to bend, curl
115.	30	*-lat-₁	15	to open the eyes wide
116.	30	*-Neŋ	15	to stare, look fixedly
117.	30	*-ŋik	15	a shrill throaty sound

Display 2 (continued)

	OC		EIMs	
118.	30	*-ŋis-₁	15	to bare the teeth
119.	30	*-pag	15	to strike, beat
120.	30	*-ris-₁	15	a rustling sound
121.	30	*-Riŋ	15	to ring
122.	30	*-taw-₁	15	to float
123.	30	*-tut	15	flatulence; stench
124.	31	*-bid	14	to twist, twine together
125.	31	*-kat	14	to rise, climb
126.	31	*-puk-₃	14	to gather, flock together
127.	31	*-rud	14	to scratch, scrape
128.	31	*-tay	14	suspension bridge
129.	32	*-baj	13	to untie, unravel, unwind
130.	32	*-baw-₃	13	shallow
131.	32	*-beŋ-₁	13	to block, stop, dam
132.	32	*-buR-₃	13	to strew, sow; sprinkle
133.	32	*-det	13	packed in, compressed
134.	32	*-gaŋ	13	dry
135.	32	*-kur	13	to coo; turtledove
136.	32	*-kus	13	to wind around; bundle
137.	32	*-laC	13	to shine; flickering or flashing light
138.	32	*-law	13	dazzling light
139.	33	*-ñeb	13	to dive; sink, disappear under water
140.	32	*-ñej	13	to submerge, drown
141.	32	*-ŋeR	13	to hear; noise
142.	32	*-piq-₂	13	to fold
143.	32	*-tus	13	to break under tension
144.	33	*-bak-₁	12	peeling skin
145.	33	*-buR-₁	12	to mix together, stir
146.	33	*-buR-₄	12	turbid
147.	33	*-kek	12	to shriek, creak, cluck, chuckle
148.	33	*-kub-₂	12	to surround, encircle
149.	33	*-kud	12	cane, walking stick
150.	33	*-leb	12	to sink, disappear under water
151.	33	*-lit-₃	12	to wind, twist
152.	33	*-luR	12	to flow
153.	33	*-muk-₂	12	mosquito, sandfly
154.	33	*-pi-₁	12	dream
155.	33	*-pis-₁	12	to deflate; empty
156.	33	*-rik	12	a spot, freckle
157.	33	*-tak	12	mud, earth, ground
158.	33	*-teŋ-₃	12	to stretch; taut
159.	33	*-teq	12	sap, gummy secretion
160.	33	*-zur	12	to thrust out, protrude
161.	34	*-bak-₃	11	to split off

Display 2 (continued)

	OC		EIMs	
162.	34	*-baŋ-$_1$	11	broad, wide
163.	34	*-bir	11	rim, edge
164.	34	*-buŋ-$_1$	11	deep resounding sound
165.	34	*-buŋ-$_3$	11	roof ridge
166.	34	*-bus-$_1$	11	to end; finished, used up
167.	34	*-bus-$_2$	11	to leak, spill out, as through a hole
168.	34	*-duk	11	ladle, spoon
169.	34	*-guC	11	to pull with a jerk
170.	34	*-ked	11	to prop, support; staff
171.	34	*-kiŋ	11	a clear ringing sound
172.	34	*-kit-$_1$	11	to bite
173.	34	*-laŋ-$_1$	11	to glitter, flash
174.	34	*-lat-$_2$	11	scar
175.	34	*-muR-$_2$	11	to gargle, rinse the mouth
176.	34	*-neŋ	11	still, tranquil, calm, as water
177.	34	*-neR	11	to hear
178.	34	*-ñep	11	to dive, sink, disappear under water
179.	34	*-paq	11	to spit; thing chewed but not swallowed
180.	34	*-peŋ	11	to plug up, dam; cover
181.	34	*-pid	11	to braid, wind together
182.	34	*-put	11	to puff, blow hard
183.	34	*-ReC	11	to tighten, as a belt
184.	34	*-sep	11	to sip, suck
185.	34	*-seq	11	to wash clothes; wet
186.	34	*-zut-$_1$	11	to pinch, pick up with thumb and forefinger
187.	34	*-zut-$_2$	11	to uproot, pluck, pull out
188.	35	*-bay	10	to accompany; partner
189.	35	*-baġ	10	the sound of a heavy smack
190.	35	*-ceq	10	to hatch, break into large pieces
191.	35	*-Cem	10	dark, of color; a dark color
192.	35	*-dek-$_1$	10	to hiccough, sob
193.	35	*-gak-$_1$	10	proud; to boast
194.	35	*-gik	10	a shrill throaty sound
195.	35	*-kel-$_2$	10	harsh coughing
196.	35	*-keŋ-$_1$	10	cramps, stiffening of limbs
197.	35	*-lem-$_2$	10	in, inside; deep
198.	35	*-naw	10	an enclosed body of water
199.	35	*-nit	10	to remove, extract
200.	35	*-nut	10	husk, fiber
201.	35	*-ŋag	10	to look upward
202.	35	*-ŋiŋ	10	a shrill buzz or hum; to ring
203.	35	*-pes	10	empty, deflated
204.	35	*-rus-$_2$	10	to slip or slide off
205.	35	*-Raq	10	red

Display 2 (continued)

	OC		EIMs	
206.	35	*-tes	10	to tear up
207.	35	*-til-$_2$	10	small protruding part
208.	36	*-bek-$_2$	9	rotten, crumbling, pulverized
209.	36	*-beŋ-$_2$	9	dull, resounding sound
210.	36	*-buq-$_1$	9	to grow, increase
211.	36	*-buR-$_2$	9	rice gruel; rice porridge
212.	36	*-Ceg	9	to hit, pound, beat
213.	36	*-Cek	9	mottled pattern
214.	36	*-daŋ-$_2$	9	to warm by a fire
215.	36	*-daR	9	to lean on, recline
216.	36	*-geŋ	9	to hum, buzz
217.	36	*-gut-$_2$	9	to gnaw
218.	36	*-kas-$_1$	9	to begin
219.	36	*-kaw-$_2$	9	high, tall
220.	36	*-lak-$_2$	9	to split
221.	36	*-laŋ-$_3$	9	spotted, striped
222.	36	*-naŋ	9	to shine, sparkle
223.	36	*-ñam	9	savory, tasty
224.	36	*-ñaw-$_1$	9	to melt, liquefy
225.	36	*-ñut	9	stretchy, elastic; springy
226.	36	*-ŋaw-$_1$	9	confused, disoriented, lost
227.	36	*-ŋaw-$_4$	9	rice bug, insect destructive to crops
228.	36	*-ŋel	9	deaf
229.	36	*-ŋuk	9	a deep throaty sound
230.	36	*-qit	9	hook, barb
231.	36	*-sir	9	hissing sound
232.	36	*-taŋ	9	clanging sound
233.	36	*-tik-$_2$	9	small, little, few
234.	36	*-tuk-$_4$	9	top, summit, crown
235.	37	*-baw-$_2$	8	rat, mouse
236.	37	*-but-$_2$	8	hole
237.	37	*-Caq	8	mud, earth, ground
238.	37	*-daŋ-$_1$	8	to dazzle, shine
239.	37	*-get	8	angry
240.	37	*-ka	8	to split, force open
241.	37	*-kaŋ-$_3$	8	stiff, rigid; cramps
242.	37	*-kiq	8	a high-pitched vocal sound
243.	37	*-kis-$_1$	8	to scratch, grate, scrape
244.	37	*-kis-$_2$	8	to tie, bind, wrap around
245	37	*-laŋ-$_2$	8	intervening space
246.	37	*-laq-$_2$	8	tongue; to lick
247.	37	*-leC	8	interval, gap, intervening space
248.	37	*-lem-$_3$	8	to sink, submerge
249.	37	*-mis	8	sweet

Display 2 (continued)

	OC		EIMs	
250.	37	*-Neb	8	door; doorway
251.	37	*-NiC	8	to skin, peel off
252.	37	*-ŋab	8	to gape; open, of the mouth
253.	37	*-ŋeŋ	8	to buzz, hum
254.	37	*-pap	8	flattened
255.	37	*-pel	8	plug, stopper; to cram
256.	37	*-pil	8	to attach, join
257.	37	*-piq-$_1$	8	to break off a piece
258.	37	*-rus-$_1$	8	a rustling sound
259.	37	*-Ris	8	to drip
260.	37	*-suR	8	satiated, full after eating
261.	37	*-teb	8	to prune, graze
262.	37	*-tuq	8	to fall
263.	37	*-yuŋ	8	to shake, sway, stagger
264.	37	*-zaŋ	8	long, tall
265.	38	*-baŋ-$_2$	7	door, gate
266.	38	*-baŋ-$_3$	7	to fly
267.	38	*-baR	7	wide
268.	38	*-buq-$_2$	7	to fall
269.	38	*-but-$_1$	7	buttocks, bottom
270.	38	*-dek-$_2$	7	pulverized, pounded fine
271.	38	*-deŋ	7	straight; to straighten
272.	38	*-dik	7	small
273.	38	*-laj	7	to spread out to dry in the sun
274.	38	*-lep-$_2$	7	to sink, submerge
275.	38	*-liŋ-$_2$	7	cross-eyed
276.	38	*-liR	7	to flow
277.	38	*-mes	7	to squeeze, knead
278.	38	*-mul	7	to suck on s.t.
279.	38	*-ŋuŋ	7	deep buzzling or humming
280.	38	*-ŋur	7	a low-pitched sound
281.	38	*-pet-$_1$	7	firefly
282.	38	*-pul	7	blunt, dull
283.	38	*-puŋ-$_2$	7	to float
284.	38	*-pus-$_2$	7	the sound of escaping air
285.	38	*-Raŋ-$_2$	7	clanging sound
286.	38	*-sit	7	to hiss, sizzle
287.	38	*-tay-$_1$	7	to hang
288.	38	*-tek-$_2$	7	mud
289.	38	*-teŋ-$_1$	7	to hum, drone
290.	38	*-tuk-$_1$	7	beak
291.	39	*-bis	6	to drip
292.	39	*-bit-$_2$	6	a whip; to whip
293.	39	*-bur	6	to stir, mix

Display 2 (continued)

	OC		EIMs	
294.	39	*-but-$_3$	6	husk, coarse hair or fiber
295.	39	*-duR	6	thunder; the rumbling of thunder
296.	39	*-gaw	6	confused, disoriented, lost
297.	39	*-ger	6	to shake, shiver, tremble
298.	39	*-kab-$_2$	6	to snap at with the teeth
299.	39	*-kad	6	to prop, support
300.	39	*-kaŋ-$_1$	6	to bark, croak
301.	39	*-keŋ-$_2$	6	hollow, resounding sound
302.	39	*-keR	6	stiff, rigid, as a corpse
303.	39	*-kit-$_4$	6	small
304.	39	*-kum	6	to close by folding
305.	39	*-led	6	to sink
306.	39	*-len	6	to swallow
307.	39	*-leŋ	6	sound that stuns a person
308.	39	*-lep-$_1$	6	shiny, bright
309.	39	*-liŋ-$_3$	6	to tilt to the side
310.	39	*-liq	6	to return, restore
311.	39	*-muk-$_1$	6	to crush, pulverize; powder
312.	39	*-nek	6	sound, of sleep
313.	39	*-niŋ	6	clear, limpid, of water
314.	39	*-ñet	6	stretchy, elastic
315.	39	*-ŋaw-$_2$	6	fly, generic for housefly and others
316.	39	*-ŋuy	6	to swim
317.	39	*-pak-$_2$	6	to peel bark from a tree
318.	39	*-paŋ	6	lame
319	39	*-paw	6	exceed, surpass
320.	39	*-pay-$_1$	6	to drape over, hang down
321.	39	*-pay-$_2$	6	to wave, flap
322.	39	*-peg	6	to hit, beat
323.	39	*-pen	6	tooth
324.	39	*-pi-$_2$	6	to fold
325.	39	*-pit-$_1$	6	near, close to
326.	39	*-sak	6	ripe, cooked
327.	39	*-sem	6	sour
328.	39	*-suŋ	6	rice mortar
329.	39	*-teŋ-$_2$	6	to stare, look fixedly
330.	39	*-ter	6	to shiver, tremble
331.	39	*-tik-$_1$	6	a gentle curve
332.	39	*-tuk-$_3$	6	to nod
333.	40	*-ban	5	group, company
334.	40	*-beg	5	thump, dull sound of collision
335.	40	*-bet	5	buttocks
336.	40	*-bun-$_1$	5	drizzle, mist, fog; hazy vision
337.	40	*-cak-$_2$	5	to stab

Display 2 (continued)

	OC		EIMs	
338.	40	*-cek	5	blind
339.	40	*-Ceŋ	5	stinging nettle: *Laportea* spp.
340.	40	*-git	5	anger, resentment
341.	40	*-huR	5	to mix, as food with water
342.	40	*-kaw-$_1$	5	a curve; to curve; circuitous
343.	40	*-kem-$_2$	5	to lie face down, be prone
344.	40	*-keŋ-$_3$	5	to shrink, shrivel
345.	40	*-kit-$_3$	5	to remove, detach
346.	40	*-lut	5	to twist around
347.	40	*-mak	5	mat
348.	40	*-mit	5	small, slight
349.	40	*-muR-$_1$	5	dew
350.	40	*-nab	5	to flood, inundate
351.	40	*-ñit	5	stretchy, elastic
352.	40	*-Naw	5	clear, pure, of water
353.	40	*-Nuŋ	5	to stare, look fixedly
354.	40	*-ŋar	5	to howl, shout, scream
355.	40	*-ŋaw-$_3$	5	leaking air, vapor
356.	40	*-ŋek	5	to grunt, groan
357.	40	*-ŋis-$_2$	5	cruel
358..	40	*-ŋit	5	to cry out, shriek
359.	40	*-paD	5	flat, level
360.	40	*-ped	5	to block, obstruct; constipated
361.	40	*-pet-$_2$	5	to hold, grasp, catch
362.	40	*-pin	5	protective later of cloth or clothing
363.	40	*-puq	5	brittle
364.	40	*-raŋ	5	bright, of light
365.	40	*-ris-$_3$	5	a slice; to slice
366.	40	*-Rud	5	to scrape
367.	40	*-sed	5	to dive, plunge into water
368.	40	*-suD	5	to budge, move a bit
369.	40	*-taq	5	raw, green, of plants
370.	40	*-taR	5	flat, level
371.	40	*-taw-$_2$	5	lost, astray
372.	40	*-tip	5	tongs; to pinch together
373.	40	*-zek	5	to step, tread, trample
374.	40	*-zel	5	dull, blunt
375.	40	*zem	5	to close the eyes
376.	40	*-zep	5	to blink; flicker, of light
377.	40	*-zul	5	to thrust out, protrude
378.	41	*-bak-$_4$	4	valley, ravine, gully
379.	41	*-bud	4	to sow, scatter seed in planting
380.	41	*-buŋ-$_2$	4	proud, haughty
381.	41	*-but-$_4$	4	unclear, hazy, misty

Display 2 (continued)

	OC		EIMs	
382.	41	*-diR	4	to lean against
383.	41	*-du	4	ladle, spoon
384.	41	*-gis	4	to scrape
385.	41	*-gut-$_1$	4	angry, annoyed
386.	41	*-hak	4	clearing of throat
387.	41	*-kul-$_2$	4	snail
388.	41	*-leR	4	defective vision
389.	41	*-lip	4	to insert, slip in
390.	41	*-lit-$_2$	4	difficult, complicated
391.	41	*-nem	4	cloud, cloudy
392.	41	*-ni	4	to hide, conceal oneself or s.t.
393.	41	*-ŋaC	4	anger, irritation
394.	41	*-ŋet	4	sweat, perspiration
395.	41	*-piŋ	4	next to, beside
396.	41	*-puŋ-$_3$	4	rice flour
397.	41	*-pur	4	to mix
398.	41	*-Raw	4	hoarse
399.	41	*-siŋ	4	to spin around
400.	41	*-teR	4	stiff, rigid, as a corpse
401.	41	*-til-$_1$	4	to pinch, pinch off a bit
402.	41	*-tiR	4	to tremble, quiver
403.	41	*-tul	4	to bounce
404.	41	*-yut	4	bag, carrying pouch
405.	41	*-zam	4	to borrow
406.	41	*-zeR	4	to stand erect

Supporting data for roots that appear in attested languages can be found in the sources given below. With regard to reconstructed languages the matter is somewhat more complex. The great majority of these appear, together with the supporting evidence, in the Austronesian Comparative Dictionary, or ACD, an open-access online resource (Blust and Trussel 2020). However, several dozen reconstructions that were cited in Blust (1988b) or subsequent publications were not found in the ACD, leaving me with one of three alternatives: 1. find evidence for the reconstruction (this presumably was found earlier in order for it to be proposed), 2. replace the proto-form with an etymologically equivalent form from a single attested language, or 3. delete the citation entirely. Alternative 1 turned out to be more difficult than I had expected, but in view of the fact that the collection of etymologies and proposed reconstructions stretch over more than half a century, this is perhaps not surprising. Wherever possible I adopted alternative 2 to avoid losing valid supporting evidence where a reconstruction could not presently be justified, but in a number of cases I simply dropped the citation on the expectation that if it was valid it would be rediscovered in future research. I hope that I have been reasonably successful in this effort, and that readers are not too often frustrated by seeing a reconstruction cited that they are unable to track down in the ACD. Many of those that do not appear in this source have been proposed in the course of writing this book, and hopefully will be added to the Austronesian Comparative Dictionary which has recently been resurrected, and should continue to grow.

4 Sources for the data

Abaknon (Jacobson n.d.)
Agta (Central Cagayan) (Reid 1971)
Agta (Eastern) (Nickell and Nickell 1987)
Agutaynen (Caabay, Edep, Hendrickson and Melvin 2014)
Aklanon (Zorc 1969)
Amis (Fey 1986)
Aputai (Taber 1993)
Arosi (Fox 1970)
Arta (Kimoto 2017)
Asilulu (Collins 2003)
Ayta Abellen (Stone 2007)

Bahasa Indonesia (Tim Penyusun Kamus 1989)
Balaesang (Himmelmann 2001)
Balangaw (Reid 1971)
Balinese (Barber 1979)
Banggai (van den Bergh 1953)
Banjarese (Hapip 1977)
Bare'e (Adriani 1928)
Batak (Dairi-Pakpak) = DPB = Dairi-Pakpak Batak (Manik 1977)
Batak (Karo) = Karo Batak (Neumann 1951)
Batak (Palawan) (Warren 1959)
Batak (Toba) = Toba Batak (Warneck 1977)
Berawan (Long Terawan) (Blust n.d-a)
Bidayuh (Bukar-Sadong) (Nais 1988)
Bikol = Naga Bikol (Mintz and Britanico 1985)
Bilaan (Koronadal) (Reid 1971)
Bilaan (Sarangani) (Reid 1971)
Bimanese (Blust n.d.-c)
Bintulu (Blust n.d.-a)
Binukid (Post and Gardner 1992)
Bisaya (Limbang) (Blust n.d.-a)
Bisaya (Sabah) (Lobel 2016)
BM = Bolaang Mongondow (Dunnebier 1951)
Bolinao (McFarland 1977)
Bontok (Reid 1976)
Buginese (Said 1977)

Bugotu (Ivens 1940)
Bukat (Diffloth n.d.)
Buli (Maan 1940)
Bulusu (Lobel 2016)
Bungku (Mead 1998)
Bunun (Blust n.d.-d)
Buruese (Devin n.d.)

Cebuano (Wolff 1972)
Chamorro (Topping, Ogo and Dungca 1975)

Dampelas (Himmelmann 2001)
Dobel (Hughes 1995)
Dumagat (Casiguran) (Headland and Headland 1974)
Dumpas (Lobel 2016)
Dusun (Kadazan) (Jekop, Topin, and Lasimbang 1995)
Dusun (Rungus) (Appell and Appell 1961)
Dusun Deyah (Hudson 1967)
Dusun Malang (Hudson 1967)
Dusun Witu (Hudson 1967)

Erai (de Josselin de Jong 1947)

Favorlang (Ogawa 2003)
Fijian (Capell 1968)

Gaddang (McFarland 1977)
Gana (Lobel 2016)
Gayo (Melalatoa et al. 1985)
Gilbertese (Sabatier 1971)
Gitua (Lincoln 1977)
Gorontalo (Pateda 1977)

Hanunóo (Conklin 1953)
Hawaiian (Pukui and Elbert 1971)
Hawu (Wijngaarden 1896)
Hiligaynon (Motus 1971)

Ibaloy (Ballard 2011)
Iban (Richards 1981)

Ibanag (McFarland 1977)
Ibatan (Maree and Tomas 2012)
Ida'an Begak (Goudswaard 2005)
Ifugaw (Lambrecht 1978)
Ifugaw (Batad) (Newell 1993)
Ilokano (Rubino 2000)
Isneg (Vanoverbergh 1972)
Itawis (Tharp and Natividad 1976)
Itbayaten (Yamada 2002)
Itneg (McFarland 1977)
Itneg (Binongan) (Reid 1971)
Ivatan (Tsuchida, Yamada and Moriguchi 1987)

Javanese (Horne 1974, Pigeaud 1938)

Kadazan Membakut (Lobel 2016)
Kalagan (Reid 1971)
Kamarian (Stresemann 1927)
Kambera (Onvlee 1984)
Kankanaey (Vanoverbergh 1933)
Kanowit (Strong n.d.)
Kapampangan (Forman 1971)
Karo Batak (Neumann 1951)
Kavalan (Li and Tsuchida 2006)
Kayan (Southwell 1980)
Kayan (Busang) (Barth 1910)
Kayan (Long Atip) (Blust n.d.-a)
Kayan (Uma Juman) (Blust n.d.-a)
Kelabit (Blust n.d.-a)
Keley-i (Hohulin, Hohulin and Maddawat 2018)
Kenyah (Long Anap) (Blust n.d.-a)
Kenyah (Long Atun) (Blust n.d.-a)
Kenyah (Long Moh) (Galvin 1967)
Kenyah (Long Wat) (Blust n.d.-a)
Kiput (Blust n.d.-a)
Komodo (Verheijen 1982)

Label (Peekel 1929–30)
Lalakai = Nakanai (Chowning n.d.-a)
Lamaholot (Pampus 1999)

Lampung (Walker 1976)
Lau (Fox 1974)
Lauje (Himmelmann 2001)
Lotud (Lobel 2016)
Lun Dayeh (Ganang, Crain and Pearson-Rounds 2006)

Ma'anyan (Hudson 1967)
Madurese (Safioedin 1977)
Makassarese (Cense 1979)
Malagasy (Richardson 1885)
Malay (Wilkinson 1959)
Maloh (King 1976)
Mandar (Muthalib 1977)
Mandaya (Barnard and Forster 1954)
Manggarai (Verheijen 1967–70)
Manggarai (West) (Verheijen 1967–70)
Manobo (Ata) (Reid 1971)
Manobo (Ilianen) (Reid 1971)
Manobo (KC) = Kalamansig Cotabato Manobo (Reid 1971)
Manobo (Sarangani) (Dubois 1976)
Manobo (Tigwa) (Reid 1971)
Mansaka (Svelmoe and Svelmoe 1990)
Mapun (Collins, Collins and Hashim 2001)
Maranao (McKaughan and Macaraya 1967)
Masbatenyo (Wolfenden 2001)
Melanau (Balingian) (Blust n.d.-a)
Melanau (Matu) (Blust n.d.-a)
Melanau (Mukah) (Blust n.d.-a)
Mentawai (Morris, 1900)
Minangkabau (Wilkinson 1959)
Miri (Blust n.d.-1971)
Molbog (Lobel 2016)
Molima (Chowning n.d.)
Mono-Alu (Wheeler 1926)
Mota (Codrington and Palmer 1896)
Motu (Lister-Turner and Clark 1930)
Muna (van den Berg 1996)
Murik (Blust 1974)
Murut, Timugon (Brewis and Brewis 2004)

Ngadha (Arndt 1961)
Ngaju Dayak (Hardeland 1859)
Nggela (Fox 1955)
Nias (Sundermann 1905)
Niue (McEwen 1970)

Old Javanese (Zoetmulder 1982)

Paiwan (Ferrell 1982)
Palauan (McManus 1977)
Palawano (Macdonald 2011)
PAN = Proto-Austronesian (Blust and Trussel 2020)
Pangasinan (Benton 1971)
Paulohi (Stresemann 1927)
Pazeh (Li and Tsuchida 2001)
PBT = Proto-Bungku-Tolaki (Mead 1998)
PCEMP = Proto-Central-Eastern Malayo-Polynesian (Blust and Trussel 2020)
PCMP = Proto-Central Malayo-Polynesian (Blust and Trussel 2020)
PEG = Proto-East Gorontalic (Usup 1981)
PMic = Proto-Micronesian (Bender et al. 2003)
PMP = Proto-Malayo-Polynesian (Blust and Trussel 2020)
PNV = Proto-North Vanuatu (Clark 2009)
POC = Proto-Oceanic (Blust and Trussel 2020)
PPH = Proto-Philippines (Blust and Trussel 2020)
PR = Proto-Rukai (Li 1977)
PS = Proto-Sangiric (Sneddon 1984)
PSS = Proto-South Sulawesi (Mills 1975)
Puyuma (Nanwang) (Cauquelin 2015)
Puyuma (Tamalakaw) (Tsuchida 1980)
PWMP = Proto-Western Malayo-Polynesian (Blust and Trussel 2020)

Rejang (Jaspan 1984)
Rembong (Verheijen 1977)
Romblomanon (Newell 2006)
Rotinese (Jonker 1908)
Roviana (Waterhouse 1949)

Sa'a (Ivens 1929)
Sa'ban (Blust n.d.-a)
Saisiyat (Li 1978)

Sambal (Botolan) (Houck and Quinsay 1968)
Samoan (Milner 1966)
Sangir (Steller and Aebersold 1959)
Sasak (Goris 1938)
Seediq (Truku) (Pecoraro 1977)
Seimat (Blust n.d.-b)
Selaru (Drabbe 1932b)
Sika (Meyer 1937)
Simalur (Kähler 1961)
Singhi (Reijffert 1956)
Soboyo (Fortgens 1921)
Subanen (Sindangan) = Central Subanen (Reid 1971)
Subanon = Siocon Subanen (Reid 1971)
Sundanese (Coolsma 1930)

Tae' (van der Veen 1940)
Tagalog (English 1986)
Tagbanwa (Aborlan) (Reid 1971)
Tagbanwa (Central) (Scebold 2003)
Tagbanwa (Kalamian) (Reid 1971)
Tarakan (Beech 1908)
Tausug (Hassan, Ashley and Ashley 1994)
Tboli (Awed, Underwood and van Wynen 2004)
Teluti (Stresemann 1927)
Tetun (Morris 1984)
Thao (Blust 2003b)
Tidung (Kalabakan) (Lobel 2016)
Tidung (Malinau) (Lobel 2016)
Tidung (Sumbol) (Lobel 2016)
Tingalan (Lobel 2016)
Tiruray = Teduray (Schlegel 1971)
Tolai (Lanyon-Orgill 1962)
Tombonuwo (King and King 1990)
Tongan (Churchward 1959)
Tontemboan (Schwarz 1908)
Totoli (Himmelmann 2001)
Tunjung (Hudson 1967)

Uma (Esser 1964)

Vitu (van den Berg 2008)

Waray-Waray (Abuyen 2000)
WBM = Manobo (Western Bukidnon) (Elkins 1968)
Wolio (Anceaux 1987)

Yakan (Behrens 2002)
Yamdena (Drabbe 1932a)
Yami (Rau, Dong and Chang 2012)
Yogad (McFarland 1977)

5 The Data

The roots identified to date are given below with full supporting evidence. Proto-forms are given first, followed by morpheme isolates, with each category in alphabetical order except that glottal stop occupies the position of 'q' (i.e. between *p*- and *r*-), and schwa (represented by /e/ where this is the convention in the source used, but by /ə/ in sources that distinguish it orthographically from /e/) is alphabetized like e. Most root doublets are cross-referenced, and with rare exceptions I cite base forms without affixes even though in normal sentence context an affixed form would be used.

The *b-roots

*-bag		the sound of a heavy smack (dbl. *-beg, *-bak.₂)
1. Ida'an Begak	agbag	to strike with a club
2. Masbatenyo	mag-balbág	to hit, pound, strike
3. Maranao	barobag	to knock, beat
4. BM	bobag	to hit with s.t. long, as a stick
5. WBM	rumbag	to throw or hit s.t. against s.t. else
6. Binukid	sumbag	to hit s.o. with the fist, to punch
7. Ida'an Begak	tərubag	the sound of marbles beating against each other in a tin
8. Sundanese	tɨmbag	bump or slam against s.t. with force
9. Tiruray	timbag	to bump s.t.
10. Ida'an Begak	tubag	to bang one's head against s.t.

*-baj		to untie, unravel, unwind
1. PWMP	*bajbaj	to loosen, untie, unwrap; clear forest
2. PPH	*hubáj	to untie, unravel
3. PMP	*lubaj	to untie, unfasten, take off
4. Ngadha	bheva	to loosen, untie
5. Buruese	hafa-h	to unravel, untie, unwind
6. Buruese	hufa-h	to undo, untie, take off
7. Cebuano	húlbad	to undo s.t. tied, untie
8. Arosi	iha	to unwrap, as food cooked in leaves
9. DPB	kimbaŋ	to open a book or letter
10. Buruese	lafa-h	to remove (a cover), open (a book)
11. Ngadha	ova	to loosen, untie, unbutton
12. Agutaynen	tolbad	to untie s.t.; to undo stitching
13. Ngadha	tova	to loosen, untie

https://doi.org/10.1515/9783110781694-006

*-bak.₁		peeling skin or bark (dbl. *-pak)
1. PMP	*bakbak	to peel off, remove bark
2. PWMP	*Rabak	to tear; in tatters
3. PWMP	*ubak	tree bark
4. Tetun	abak	peeling (a disease of the skin)
5. Kapampangan	balikúbak	dandruff
6. Javanese	bəbak	to peel the husk or skin from
7. Iban	kərəbak	lift off, peel open (as paint that is coming off)
8. Muna	obha	remove the outer layer (of a piece of bark)
9. Palauan	omkóbk	to peel (banana, etc.); peel off (bark, skin after sunburn, etc.)
10. Bikol	sabák	to remove the bark of the *danlóg* tree
11. Keley-i	təbak	to remove the bark of a tree
12. Rejang	təkəluba?	scratched, skin torn

*-bak.₂		the sound of a heavy smack (dbl. *-bek.₂, *-buk.₂, *-bag)
1. PMP	*ambak	to stamp or smack against
2. PMP	*bak	clap! smack!
3. PAN	*bakbak	sound of heavy clapping or pounding
4. PMP	*be(m)bak	to slap, beat on
5. PAN	*Cebak	to smack against
6. PWMP	*debak	to pound, thud
7. PWMP	*kibak	to flap, clash together
8. PMP	*la(m)bak	to slam s.t. down
9. PWMP	*lebak	to pound, thud
10. PMP	*sambak	to hit, strike a heavy blow
11. PMP	*ta(m)bak	to hit, pound on
12. PWMP	*ti(m)bak	to clap, make a clapping sound
13. PMP	*tu(m)bak	to clash together
14. Yakan	balakbak	the sound of gurgling or bubbling liquid
15. Ngaju Dayak	bikbak	plop of fruit falling on soft ground
16. DPB	bukbak	pounding of the heart, as in fear
17. Malay	cəbak-cəbok	to go splashing through slush
18. Iban	dumbak	cylindrical drum with skin at each end
19. DPB	émbak	work the soil with a hoe
20. DPB	gəbak-gəbak	sound of falling objects
21. Bikol	gúbak	to beat or punch
22. Javanese	kə-kəbak	beating of wings (Pigeaud 1938)
23. Bikol	lubák	to pound (bananas, root crops)
24. Singhi	pabak	hammer

(continued)

25. Tagbanwa (Central)	rabak	the sound of rain or tapping or rapid gunfire
26. Mandar	rimbaʔ	beat the wings
27. Maranao	robak	to hammer; sledgehammer
28. Agutaynen	sigbak	to beat or pound on a person
29. Sundanese	təbak	to slam against with force, as breakers against the rocks
30. Lamaholot	təribaʔ	to hit sideways with the hand
31. Tagbanwa (Central)	tigbak	to hit s.o. on the back

*-bak₃		to split off (dbl. *pak₁)
1. PMP	*ibak	to break or split off
2. PWMP	*sibak	to split open
3. Itbayaten	axbak	splitting or chopping wood lengthwise
4. Ilokano	gíbak	potsherd, fragment of broken pottery thinner than a *ríbak*
5. Iban	kəlibak	to chip with axe or knife
6. Ilokano	ríbak	potsherd, fragment of broken pottery
7. Toba Batak	silbak	splinter, splintered
8. Kayan	tebak	split in two lengthwise
9. Iban	tibak	to cut a piece off
	tiba-tibak	ragged
10. Bontok	tobak	chop a slab from a log
11. Cebuano	úbak	to separate layers of banana trunk

Note: Despite a general semantic similarity, *bak₁ 'peeling skin', and *-bak₃ 'to split off' appear to be distinct roots. Cebuano *úbak* may belong to *-bak₁.

*-bak₄		valley, ravine, gully
1. PWMP	*lebak	ravine, glen, valley
2. Lun Dayeh	abak	a small gully/valley
3. Balaesang	eβak	valley
4. Tagalog	lambák	valley

*-ban		group, company
1. PMP	*baban	group,
2. PWMP	*kaban	group, flock, herd
3. Ilokano	arbán	flock, herd, drove
4. Hiligaynon	gubán	group, brigade, caravan
5. Cebuano	ubán	companion

*-baŋ.₁		broad, wide (dbl. *-baR)
1. PWMP	*baŋban	broad, spacious
2. PWMP	*lambaŋ	broad, wide
3. Dusun (Kadazan)	hivaŋ	width; broad, spacious, ample
4. Toba Batak	hombaŋ	spread out broadly
5. Dusun (Kadazan)	kivaŋ	wide, broad, spacious
6. Bikol	lakbáŋ	breadth, width
7. Balinese	lumbaŋ	broad, wide
8. Mansaka	marakbaŋ	wide
9. Kambera	rambaŋu	broad, wide
10. Sasak	rəmbaŋ	broad shouldered
11. Manggarai	ribaŋ	wide; open space

*-baŋ.₂		door, gate (dbl. *-waŋ)
1. PMP	*ebaŋ	doorway, door opening
2. Buginese	babaŋ	door
3. Malay	gərbaŋ	opening out, as a fan
4. Dusun (Kadazan)	ivaŋ	to open (of door or window)
5. Madurese	labaŋ	door
6. Manggarai	lewaŋ	door opening
7. Tarakan	tumbaŋ-an	door, gate

Note: Also Balaesang, Wolio *bamba* 'door', which agree with Buginese in pointing to *baŋban in a proto-language that is difficult to identify, and Banggai *mbamba* 'door', which points instead to a form with open final syllable. Because it has undergone the change *w > b, Madurese *labaŋ* may instead contain the root *-waŋ 'wide open space'.

*-baŋ.₃		to fly
1. Yami	alibaŋbaŋ	flying fish
2. Malay	ambaŋ	afloat in mid-air (as a bird resting on its pinions)
3. BM	bombaŋ	to fly up, rise together, as a swarm of flies or flock of chickens
4. Toba Batak	habaŋ	to fly
5. Itbayaten	liibaŋ	flying fish
6. Kanowit	səbabaŋ	butterfly
7. Malay	tərbaŋ	to fly

Note: A form matching Yami *alibaŋbaŋ* 'flying fish' means 'butterfly' in many other languages (Blust 2001). PAN *b normally became Itbayaten -*v*-; the retention of a stop in this form is unexplained.

*-baR		wide (dbl. *-baŋ₋₁)
1. Binukid	balabag	width; to lie across s.t.
2. Manggarai	jimbar	wide and large
3. Paiwan	me-lava	to be wide
4. Ngaju Dayak	lumbah	wide, as a path or road
5. Sarangani Bilaan	mabal	wide
6. Isneg	pabbág	the breadth or width of the house, at the outside
7. Maloh	ribar	wide, extensive

Note: Also Balinese *jimbar* 'broad, extensive, wide', Rejang *umbar* 'wide, spacious'.

*-baw₋₁		high; upper surface (dbl. *-kaw₋₂)
1. PMP	*abaw	high, lofty
2. PAN	*babaw	upper surface
3. PMP	*bawbaw	upper surface, top; above
4. PWMP	*embaw	high, on top
5. PPH	*lábaw	project above, stick up
6. PWMP	*timbaw	high up, on top
7. PMP	*umbaw	top part; high
8. Isneg	alíbaw	the tallest of all spirits
9. Maranao	gambaw	top of; overlap
10. Mansaka	gombaw	to be higher than
11. Cebuano	ígbaw	at the top of something
12. Thao	kafaw	high; at a height
13. Kenyah	kembaw	over, up, above
14. Bikol	mabáw	to be almost full (of a container)
15. Isneg	rabáw	top, peak of a mountain
16. Ilokano	rimbáw	top, peak, summit
17. Hanunoo	sakbáw	uplands, mountains
18. Kayan	selibaw	suspended above
19. Singhi	de somu	above
20. Aklanon	támbaw	be over one another, overlapping
21. Maranao	tebaw	stand out in a crowd due to height
22. Kayan	telibaw	directly above (as in bunk beds)
23. Bontok	təŋbaw	remove the top part of the contents of a container
24. Cebuano	túgbaw	lofty, high up
25. Bikol	umbáw	to be filled to the brim, full to the top
26. Cebuano	úsbaw	to rise in degree or quantity

*-baw₂		rat, mouse
1. PWMP	*babaw	rat, mouse
2. PMP	*balabaw	rat, mouse
3. PCEMP	*kalabaw	rat, mouse
4. PMP	*labaw	rat, mouse
5. Binukid	ambaw	rat, mouse
6. Paiwan	kulavaw	rat, mouse
7. Kavalan	mrabaw	a field rat
8. Aputai	palaho	mouse, rat

*-baw₃		shallow
1. PWMP	*babaw	shallow
2. Itawis	ababbáw	shallow
3. Tagalog	dábaw	fording or wading place
4. Waray-Waray	hagmabaw	shallow
5. Itbayaten	havavaw ~ hivavaw	shallowness
6. Maloh	imbaw	shallow
7. Kenyah	ləbaw	shallow
8. Kenyah	libaw	a ford; shallow
9. Cebuano	mabáw	shallow
10. Aklanon	nábaw	become shallow(er)
11. Kayan (Long Atip)	ñivaw	shallow
12. Dusun (Rungus)	o-ribaw	shallow
13. Tiruray	təmbaw	shallow

*-bay		to accompany; partner
1. PWMP	*abay	side by side
2. PAN	*sabay	to do s.t. together with others
3. Waray-Waray	agbáy	the act of putting one's arm over the shoulder of another
4. Bikol	antábay	to accompany, escort, stay with
5. Bikol	gubáy-gúbay	to be almost the same age as s.o. else
6. Waray-Waray	hagubay	the act of embracing
7. Bikol	ŋábay	to accompany s.o. on a hunt
8. Waray-Waray	rumbáy	to line up, form a line
9. Ida'an Begak	sagbay	to put an arm around s.o.'s shoulder
10. Bikol	saŋbáy	mistress, lover

Note: Also Ayta Abellen *agapay* 'companion'.

*-beg		thump, dull sound of collision (dbl. *-bag, *-peg)
1. Ibaloy	banegbeg	thud, thump, bump—the noise of impact, esp. foot sounds (as stomping, footsteps)
2. Maranao	barobəg	to whip, beat
3. Dumagat (Casiguran)	bəlbə́g	to beat up someone
4. Maranao	bəmbəg	to knock, beat
5. Old Javanese	gədəbəg	pounding, stamping

*-bej		to wind around repeatedly
1. PAN	*bejbej	to wind around repeatedly
2. PWMP	*embej	to bind the waist; waist band
3. PWMP	*hambej	to wrap around; band around s.t.
4. PWMP	*kabej	to tie up, tie by wrapping around
5. PPH	*la(m)bej	to wrap or wind around s.t.
6. PWMP	*libej	to coil around, wrap with rope
7. PWMP	*qambej	to wind around
8. PWMP	*Ra(m)bej	to wind around
9. PWMP	*ta(m)bej	to tie up, bind tightly
10. Maranao	babəd	band, bandage, tie, belt
11. Cebuano	balíbud	to wind s.t. around s.t.
12. Tagalog	balíbid	twine tied around several times
13. Aklanon	bolímbod	to tie by running the string around many times
14. Ifugaw	hibód	act of encircling s.t. with rope, etc.
15. Malay	kəbat	binding round; band; wrapper; circlet
16. Maranao	kibəd	to wind
17. WBM	kulambəd	of a vine, to climb by entwining itself on other plants or trees
18. Maranao	lobəd	to wrestle
19. Hiligaynon	sámbud	to wind, bind, roll
20. Javanese	ubəd	wrapped, swathed
21. Hanunóo	yabúd	tangle, entwining, as of vines

*-bek₋₁		dull, muffled sound
1. PWMP	*bek	sound of hacking or chopping
2. PAN	*bekbek	sound of breaking, etc.
3. PMP	*debek	to smack against
4. PPH	*e(m)bek	thudding sound
5. PWMP	*ha(m)bek	a sob, mumbled complaint
6. PPH	*lebék	pound in a mortar
7. PAN	*Nabek	breakers, surf, waves
8. PMP	*ra(m)bek	to strike, break
9. PMP	*se(m)bek	muffled sound of a gasp, etc.
10. PMP	*tebek	to tap, thump
11. Kankanaey	ʔabəbə́k	to thump, bump (of s.t. heavy that strikes the water)
12. Maranao	bambək	to knock, beat
13. Fijian	cobo	clap the hands crosswise so as to make a hollow sound
14. Ifugaw	imbók	beating of gongs while people dance
18. Kayan	kubek	whisper, undertone
19. Maranao	lobək	to trample, pound (as rice)
20. Bintulu	məmbək	to hit with a stick
21. Sasak	obək	to hit with the open hand
22. Sasak	rəəbək	to hit, strike
23. Kankanaey	sibə́k	to sob (mostly broken sobs, after crying)

*-bek₋₂		rotten, crumbling, pulverized (dbl. *-buk₋₁, *-dek₋₂, *-mek, *-pek)
1. PMP	*bekbek	dust of decaying wood
2. PPH	*lebék	to pound (rice, etc.) in a mortar
3. PPH	*lubek	to pulverize, pound into dust
4. Kankanaey	gəbə́k	rotten; with rotten haulm, stalk, culm
5. Itbayaten	koxbək	dust
6. Mansaka	lərəbək	pestle, pole for pounding grain
7. Aklanon	límbok	pounded rice
8. Ilokano	rebbek	ruins, waste; destruction, demolition
9. Kankanaey	talbə́k	beaten, etc. to pieces, pulverized
10. Amis	tifək	to mill grain

*-beŋ.₁		to block, stop, dam (dbl. *-peŋ)
1. PAN	*beŋbeŋ	blocked, as by a wall or curtain
2. PMP	*embeŋ	a dam; to dam a stream
3. PWMP	*hambeŋ	to obstruct, block the way
4. PWMP	*qambeŋ	to obstruct, block the way
5. PWMP	*tambeŋ	dam, obstruction of flow
6. PWMP	*tubeŋ	obstacle; to block, obstruct the flow
7. PMP	*ubeŋ	to block, dam up
8. Arosi	ʔabo	firm, tight, as a cork in a bottle
9. Itbayaten	rivəŋ	shield, lee, windbreak; covering
10. Bikol	simbóŋ	stopper of leaves; plug the mouth of a container with a stopper of leaves
11. Madurese	tebbeŋ	closed off with a wall
12. Tboli	g-tibeŋ	covered, as with soil; obstructed, hindered, blocked, as of light
13. Vitu	zobo	blocked (of road, pipe)

*-beŋ.₂		dull resounding sound (dbl. *-buŋ.₁)
1. PAN	*beŋbeŋ	buzz, hum
2. PWMP	*e(m)beŋ	droning or buzzing sound
3. Lun Dayeh	beŋ	the sound of an empty bottle when the cork is pulled
4. Dusun (Kadazan)	koboŋ	make a banging noise
5. Malay	kumbaŋ	carpenter-bee; bumblebee
6. Itbayaten	mabəŋ	low-pitched resounding tone
7. Iban	səbaŋ	large drum
8. Arosi	toho	groan in sickness
9. Kayan	tubeŋ	long drum made from a hollowed log

*-bet		buttocks (dbl. *-but)
1. PWMP	*lubet	buttocks
2. PMP	*ubet	buttocks, posterior
4. Bontok	kóbət	buttocks
5. Keley-i	lihbət	big buttocks

*-bid		to twist, twine together (dbl. *-pid)
1. PWMP	*bidbid	to twist, wrap around
2. PPH	*bidibid	to twist together, intertwine
3. PPH	*lambid	to cross over, wind around, embrace make rope
4. PPH	*lúbid	string, rope, cordage made by twining fibers
5. PWMP	*tabid	to twist, braid
6. Maranao	binibid	to twist, entwine
7. Bikol	biríbid	gnarled, twisted
8. Maranao	boribid	to wring, twist; crooked
9. Hanunóo	láŋbid	strand, referring to bast fibers in a twisted rope; a unit classifier for such strands
10. Mansaka	linabid	to have one's legs crossed (as when sitting)
11. Hanunóo	sálbid	crosslegged
12. Maranao	sobid	to strengthen rope by twining it with additional twine
13. Maranao	tarəmbibid	to twist; crooked; tortion
14. Kankanaey	túbid	to plait, braid, twist

*-bir		rim, edge
1. PMP	*birbir	rim, edge, border
2. PWMP	*su(m)bir	rim, border
3. PWMP	*ta(m)bir	rim
4. PPH	*te(m)bir	edge, seam, hem
5. DPB	abir	to edge away, move aside
6. Simalur	befil ~ fefil	rim, edge, side
7. Ngadha	kebi ~ kebhi	rim, edge, border
8. Javanese	mbir	edge, side of
9. Iban	pələbir	placed to hang over an edge
10. Ngadha	rubi	rim, edge, border, crest of mountains
11. Madurese	sembir	edge, border, outskirts

Note: Also *simpir 'edge, margin'.

*-bis		to drip (dbl. *bus.₂, *-Ris)
1. Bikol	b-ar-isbís	to flow or run, referring only to rain water which runs from the roof
2. Sundanese	bibis	to sprinkle with water (as tobacco leaves to keep them fresh)
3. Sundanese	biribis-biribis	drizzle, rain lightly
4. Itawis	mab-bisíbit	to water
5. Maranao	boribis	to sweat freely
6. Bikol	tibís-tibís	to drip rapidly or repeatedly

*-bit.₁		to hook, clasp; grasp with fingers (dbl. *-wit)
1. PMP	*ambit	to seize with the hands
2. PAN	*bitbit	to pull at some part of the body; hold dangling from the fingers
3. PWMP	*cibit	to hook on, catch with a hook
4. PWMP	*cubit	to take with the fingers
5. PMP	*ibit	to touch, grasp
6. PAN	*kabit	hook
7. PPH	*kabit	to lead by the hand, support (as a feeble person)
8. PPH	*kal(e)bit	to touch lightly, nudge; pull trigger
9. PMP	*kebit	to touch or tap to draw attention
10. PMP	*kubit	to touch lightly
11. PPH	*labit	to hook with the fingers
12. PWMP	*ra(m)bit	to hook together
13. PWMP	*sabit	hooked implement
14. PPH	*sak(e)bit	to hang, suspend from something
15. PMP	*saqebit	to hook onto s.t., entangle
16. PSS	*ti(bw)i(C)	to carry
17. Iban	bəribit	to carry by a fold (as in a bag or cloth), (of crabs) carry in the claws
18. Old Javanese	bibit	fishhook
19. Bahasa Indonesia	gubit	to summon by waving (fingers motioning downward and inward)
20. Dusun (Kadazan)	haŋkavit	claw, talon
21. Dusun (Kadazan)	hoŋkivit	to pull down with the hand
22. Malay	jimbit	lifting (anything) off the ground with hand or finger
23. Tagalog	kalabít	touching with the tip of the fingers; strumming of a stringed instrument
24. Dusun (Kadazan)	kibit	to carry a baby in one's arm
25. Cebuano	kulumbábit	to hang or cling onto something
26. Kenyah	ŋe-libit	to grip, hold on to
27. Tetun	nak-nabit	pliers, tongs

(continued)

28.	Kenyah	nelebit	to dangle on s.t., to hang on
29.	Dusun (Kadazan)	ovit	to carry, convey, bring, take along
30.	Bintulu	səkabit	fruiting pole with a hook at the end
31.	Cebuano	sibit	safety pin
32.	Iban	tambit	fastening; latch, button
33.	Lamaholot	təbit	pick up a small bit of fish or meat with the fingertips

*-bit.$_2$		a whip; to whip
1. PAN	*labit	a whip; to whip
2. Maranao	ambit	to whip, entangle
3. Palauan	chelébed	whip; club; bat; anything to hit with
4. Kayan	hivit	to hit, whip
5. Puyuma (Tamalakaw)	a-lvit-i	to whip
6. Javanese	sambit	to strike with a whip or missile

*-buC		to weed, pluck, pull out (dbl. *-buC, *-guC, *-NuC, *-zut)
1. PMP	*abut	to pull up, root up
2. PAN	*buCbuC	to pull up (as weeds), to pluck (as feathers)
3. PMP	*cabut	to pull out with force
4. PWMP	*ebut	to pluck, pull out
5. PMP	*ibut	to pull out, uproot
6. PWMP	*ranebut	to pluck out
7. PPH	*raʔebut	to pull out s.t. rooted in soft matter
8. PAN	*RabuC	to uproot, pull out by the roots
9. PWMP	*Ranebut	to pluck out
10. PMP	*Rebut	to pull out, extract
11. PAN	*ubuC	to pluck, pull out
12. Iban	babut	to root up, pluck, pull out, weed
13. Old Javanese	dawut	pulled out, uprooted
14. Hanunóo	galbút	pulling out abruptly, as hair or teeth
15 Ifugaw	híbut	the act of weeding
16. Wolio	hobut-i	to pull out (hair, feathers)
17. Bikol	hulbót	to pull out, unsheathe, extract
18. Ibaloy	Kabot	to pull up a plant by the roots
19. Manggarai	kebut	to pluck, pull out
20. Banggai	kubut	to pull out, extract
21. Erai	lahuk	to pick, pluck off
22. Gorontalo	pahuto	to pull out, extract

(continued)

23. Manggarai	robut	to pluck (feathers) with the fingers
24. Ma'anyan	rubut	to pluck, pull out
25. Cebuano	sákbut	to snatch, grab something from s.o.
26. Seimat	sohut-i	to pluck, pull out

Note: Source for Ma'anyan *rubut* unknown.

*-bud		to sow, scatter seed in planting (dbl.*-buR.$_3$)
1. PMP	*budbud	to sprinkle seed, sow seed in dibble holes
2. PWMP	*qabud	to strew, scatter, sprinkle
3. PWMP	*sabud	to sow, strew, scatter
4. PWMP	*tabud	to strew, scatter

Note: *sabud and *tabud were earlier reconstructed as *sabuD and *tabuD respectively, but this now appears to have been because of misassignment of reflexes of the doublets *sabuR and *tabuR.

*-buk.$_1$		to decay, crumble; powder (dbl. *-bek, *puk.$_2$)
1. PPH	*alikabuk	dust
2. PWMP	*bukbuk	wood weevil
3. PMP	*dabuk	dust, ashes
4. PMP	*ebuk	dust, powder
5. PMP	*habuk	powdery substance, dust
6. PMP	*la(m)buk	rice bran or dust left from pounding
7. PMP	*qa(R)buk	dust
8. PMP	*rebuk	dry rot in wood
9. PMP	*Rabuk	dust, powder
10. PAN	*tibuk	to pound; throb
11. Tae'	barubuk	dust
12. Javanese	blubuk	(of soil, ash) soft, crumbling
13. Mandaya	gəbuk	rotten
14. Bikol	gubók	rotten (wood)
15. Ifugaw	ibuk	black and rotten (teeth)
16. Bimanese	kaluɓu	dust, ashes
17. Manggarai	kebok	dust, ashes
18. Lamaholot	kəʔawuk	ashes, dust
19. Lamaholot	məwuk	dust, powder, flour
20. Toba Batak	orbuk	dust
21. Gorontalo	peyahuʔo	dust
22. Malay	sərbok	fine particles, powder (of wood)
23. Balinese	srabuk	powder, dust
24. Kavalan	tabuk	dust as gathered in a house

(continued)

25. Arosi	tahu	bore holes; wood-boring insect
26. Sasak	təbuk	finely ground coffee
27. Bugotu	uvu	powdery, friable, floury
28. Hiligaynon	yábʔuk	dust, dirt

*-buk$_2$		to pound; a thud, heavy splash
1. PMP	*bukbuk$_1$	bubbling, boiling
2. PMP	*bukbuk$_2$	to hit, pound, beat
3. PAN	*Cebuk	thud
4. PMP	*dabuk	to pound, beat
5. PMP	*ibuk	dull, muffled sound
6. PMP	*lambuk	to knock, pound, beat
7. PWMP	*lebuk	to pound, thud
8. PMP	*lubuk	to pound with a pestle
9. PMP	*ra(m)buk	to knock, pound, beat
10. PWMP	*ribuk	noise, tumult
11. PWMP	*tebuk	to pound, as rice
12. PAN	*ti(m)buk	to pound, throb
13. PMP	*tumbuk	to punch, hit, pound
14. Puyuma	alvuk	to hit with a large stick
15. Javanese	babuk	to butt with horns
16. Tagalog	bulbók	bubbling or gushing of liquids
17. Fijian	cevu	to explode, as breadfruit in fire
18. Malay	dəbuk	a thud
19. Malay	gəbok	striking a heavy blow with a flat object
20. Rembong	kembuk	sound of splashing about in the water
21. Kambera	kubuku	sound of s.t. splashing in the water
22. Aklanon	límbok	pounded rice
23. Iban	pərəbok	bubbling or gurgling noise
24. Kambera	póbuku	hollow sound, as gourd when struck
25. Manggarai	rembuk	to pound again (of corn flour)
26. Sasak	rubuk	to bubble (of water), to snort (as people in the water)
27. Yamdena	sambuk	to pound fine (of roasted maize)
28. Kayan	sebuk	the sound of fruit or stones falling into water
29. Kankanaey	síbok	to clack, clap
30. Ida'an Begak	sugbuk	to collide
31. Bare'e	tabu	splash, plop
32. Old Javanese	tələbuk	with a thud
33. Fijian	wabu	to beat with a stick, throw down with violence

*-bun.₁		drizzle, mist, fog; hazy vision
1. PWMP	*ambun	dew, mist, fog
2. PWMP	*qembun	dew, mist, fog
3. PWMP	*rabun	blurry vision, failing eyesight
4. PMP	*Rabun	drizzling rain, mist, fog
5. Mandar	gauŋ	mist, fog

*-bun.₂		to heap up, cover with earth; collect, gather (dbl. *-pun)
1. PWMP	*abun	to collect, gather
2. PMP	*bunbun	to fill a hole with earth; to cover with
3. PMP	*hebun	group; pile
4. PMP	*hibun	to cover with earth
5. PMP	*hubun	heap, pile, collection
6. PMP	*limbun	to heap up, pile up (as earth); dam
7. PWMP	*Rabun	to cover with earth
8. PWMP	*Ri(m)bun	to heap up
9. PMP	*ta(m)bun	to cover with earth, sand, etc.
10. PWMP	*timbun	heaped up
11. PWMP	*tu(m)bun	heap, pile; to cover up
12. Makassarese	balumbuŋ	heap up earth over something
13. Ngadha	dabu	collect, gather
14. Tetun	kabun	group (of people), grove (of trees)
15. Kenyah	kelimbun	to assemble
16. Hiligaynon	kibún	to surround, besiege
17. Iban	kumbun	to wrap or cover, as with a blanket
18. Mapun	lambun	to gather into a pile
19. Kenyah	lembun	a barn, rice store; a gathering, herd, throng
20. Buruese	lubu-	a heap; in heaps or piles
21. Bimanese	ŋumbu	heaped up
22. Kenyah	pembun	to gather together
23. Ida'an Begak	rigbun	to cover with a blanket
24. Malay	rumbun	to pile wood on a fire so as to increase the blaze
25. Kenyah	selimbun	to gather together
26. Isneg	tabbón	to gather, collect, come together
27. Ida'an Begak	tagbun	to cover s.t. drying in the sun
28. Cebuano	úgbun	mound, esp. a heap of finely-textured things
29. Nggela	uruvu	to assemble

*-buŋ₁		deep resounding sound (dbl. *-beŋ₂)
1. PMP	*buŋ	deep resounding sound
2. PPH	*buŋbuŋ	deep resounding sound
3. Kapampangan	akbún	explosion
4. Hanunóo	búmbuŋ	beating on gongs
5. Kankanaey	gíbuŋ	to resound; applied to the barking of many dogs
6. Mapun	kabbuŋ	a popping sound (as when pulling a cork from a bottle)
7. Hanunóo	lágbuŋ	sound made by a stone falling into water
8. Banggai	ndibuŋ	make a plopping sound in the water
9. Balinese	tambuŋ	to echo, resound; an echo
10. Puyuma (Tamalakaw)	tevuŋ	dull sound of dropping into water, as a stone
11. Kayan	tubuŋ ~ tuvuŋ	a drum made from a long, hollow, wooden cylinder covered with the skin of an animal

Note: PPH *buŋbuŋ 'deep resounding sound', and Hanunóo búmbuŋ 'beating on gongs' are treated as EIMs since Hanunóo, preserves -ŋC- in other reconstructions, as PAN *daŋdaŋ 'to warm oneself or s.t. near a fire'; > daŋdán 'roasting directly on coals, as of bananas'.

*-buŋ₂		proud, haughty
1. PWMP	*ambuŋ	proud, haughty
2. Ayta Abellen	lamboŋ	to make a boast, to be proud of s.t.
3. Malay	somboŋ	arrogance, self-assertion; be puffed-up with pride
4. Malay	tamboŋ	stubborn, headstrong, arrogant

Note: Also PPH *hambúg 'proud, boastful'.

*-buŋ₃		roof ridge
1. PMP	*bubuŋ(-an)	ridge of the roof
2. PPH	*kubuŋ	protective covering or enclosure
3. PWMP	*kulubuŋ	head covering for women
4. PMP	*ubuŋ	ridge of the roof
5. Ngaju Dayak	babuŋ-an	ridgepole cover
6. Mandar	balimbuŋ-aŋ	ridgepole cover
7. Kenyah	beluboŋ	the house top
8. Iban	raboŋ	highest part, summit, roof ridge

(continued)

9. Puyuma (Tamalakaw)	raHuvuŋ	ridgepole of roof
10. Bidayuh (Bukar-Sadong)	tibuŋ	top; a peak, high vantage point
11. Uma	wumu	ridge of the roof

*-buq₋₁		to grow, increase
1. PPH	*talúbuq	to grow rapidly, thrive, be healthy
2. PMP	*tu(m)buq	to grow, germinate, sprout
3. Itbayaten	harovo	growth
4. Old Javanese	imbuh	increase, addition
5. Bikol	núboʔ	to develop a new breed or strain from
6. Bintulu	subuʔ	young bamboo shoot
7. Bidayuh (Bukar-Sadong)	timbu	to sprout up (plants, hair, teeth, seeds)
8. Tausug	tuŋbuʔ	the growth of plants
9. Manggarai	ubu	age; to grow

*-buq₋₂		to fall (dbl. *-tuq)
1. PWMP	*dabuq	to fall
2. PMP	*labuq	to fall; drop anchor
3. PMP	*nabuq	to fall
4. PWMP	*sabuq	to drop, fall
5. Arosi	ahu	to fall, come down
6. Yamdena	fufu	to fall from a height
7. Malay	roboh	heavy fall; crash (of the fall of buildings; tiled roofs; castle gates; great forest trees, etc.)

*-bur		to stir, mix (dbl. *-buR, *-huR, *-pur)
1. PMP	*kambur	to mix
2. PMP	*rembur	to mix together
3. Javanese	bawur	mixed (together)
4. Lun Dayeh	rabur	mixing liquid
5. Maranao	sambor	to mix, mixture
6. Isneg	tambúr	to mix shredded coconut meat with *bási* (rum)

*-buR.₁		to mix together, stir (dbl. *-bur, *-huR, *-pur)
1. PPH	*kibuR	to mix two things together
2. PWMP	*labuR	to mix foods
3. Ngadha	kabhu	to mix
4. Kambera	kaŋaburu	to mix
5. Maranao	kolibog	to stir (in cooking)
6. Bare'e	lowu	to join, mix
7. Mansaka	lambog	to mix different kinds
8. Hanunóo	líbug	a mixture
9. Hanunóo	palbúg	mixing of rice with broth; the resulting mixture
10. Cebuano	sámbug	to mix things together; ingredient; mixture of rice and corn
11. Maranao	səmbog	to mix
12. Hiligaynon	símbug	to mix, dilute liquids

*-buR.₂		rice gruel; rice porridge
1. PMP	*buRbuR	porridge
2. Kiput	bəlasauʔ	porridge (< *bəlabʰuR)
3. Mansaka	bibog	soft-cooked rice
4. Maranao	biobog	porridge
5. Binukid	bulbug	to make rice or corn porridge
6. Tiruray	ʔəbor	broth; to mix *ebor* with rice
7. Tae'	kalemboʔ	porridge, gruel
8. Hanunoo	lábug	rice gruel
9. Kelabit	menebʰur	rice gruel

Note: Also Bintulu *avud* 'porridge of sago pearls boiled with coconut meat'. This root and *-buR.₁ 'to mix together' may be the same, since rice gruel is made by mixing a liquid with cooked rice

*-buR.₃		to strew, sow; sprinkle (dbl. *-bud)
1. PMP	*ebuR	to sow, strew, scatter
2. PWMP	*imbuR	to scatter, strew
3. PMP	*kambuR	to sprinkle, scatter (seed, etc.)
4. PWMP	*qambuR	to strew, scatter, sprinkle
5. PMP	*saq(e)buR	to sow, scatter; scattered about
6. PMP	*simbuR	to sprinkle
7. PMP	*tabuR	to strew, scatter, as seed
8. Sundanese	bəlabur	to spread, scatter, disperse
9. Palauan	melúbs	to sprinkle, spray, throw water on

(continued)

10. Toba Batak	rabur	scattered, strewn
11. Malay	səmbur	bespattering from the mouth
12. Rotinese	sofu	to strew, scatter (as powder)
13. Bare'e	webu	to sprinkle

Note: Also Iban *pambar* 'scattered, dispersed'.

*-buR.₄		turbid; unclear, as water
1. PMP	*balabuR	blurred, of vision
2. PPH	*kabuR	be agitated, as water when fish swarm
3. PMP	*kebuR	to churn (water agitated by fish)
4. PWMP	*lebuR	turbid, murky
5. PMP	*libuR	murky, clouded, turbid
6. Javanese	bawur	(of vision) blurred, hazy
7. Nggela	gagavu	dirty, of water
8. Karo Batak	gəmbur	some turbid (water, urine, etc.)
9. Ifugaw	kanibul	idea of being turbid, muddy, as the water of a river after a heavy rain
10. Keley-i	kibul	the muddy or cloudy condition of a body of water that has been disturbed by a person or animal moving through it
11. Manggarai	tebur	turbid (of water)
12. Fijian	vuvu	muddy, troubled (of water)

Note: Also Iban *amur* 'mud, dust; muddy, cloudy'.

*-bus.₁		to end; finished, used up (dbl. *-pus)
1. PWMP	*ebus	finished, gone
2. PMP	*kabus	to run out, as of supplies; be poor
3. PPH	*lubús	completely, to the end
4. PWMP	*qubus	finished, gone
5. Ilokano	búbus	to consume, use up
6. Ilokano	gíbus	end, limit, finish, completion
7. Ilokano	íbus	to finish, use up, exhaust
8. Iban	kələbus	to finish off (as in doing work)
9. BM	kobut	dead (animals)
10. Yakan	tambus	to be finished, completed
11. Dusun (Kadazan)	tuvus	to finish, exhaust, use up

*-bus₂		to leak, spill out, as through a hole (dbl, *-bis, *-Ris)
1. PMP	*busbus	to leak, spill through
2. Ida'an Begak	bərubus	to fall (of objects), leak
3. Iban	bumbus	holed, pierced, make a hole in
4. Cebuano	híbus	for a sack to break and lose its contents
5. Iban	kəbus	having a hole
6. Iban	ləbus	to slip away, push through and out (as of a forest)
7. Ilokano	okbós	spilled out
8. Iban	pambus	come out or shoot out; punctured, burst
9. Ida'an Begak	tambus	pierced
10. Malay	təmbus	perforated; having a big hole knocked through it
11. Keley-i	ubuh	to get out of an enclosure through a gap or hole, as pigs escaping from a pigpen

*-but₁		buttocks, bottom (dbl. *-bet)
1. PWMP	*abut	buttocks
2. Tiruray	bubut	base, buttocks
3. Palawano	embut	anus
4. Melanau (Mukah)	kabut	buttocks, posterior; bottom, base
5. Manobo (Tigwa)	lebut	buttocks
6. Kayan	lekbut	anus
7. Bikol	lusbót	buttocks (said in jest or anger)

*-but₂		hole
1. PMP	*butbut	hole, perforation, pit
2. PPH	*labút	a hole, pit
3. PPH	*lubut	a hole; to make a hole in, perforate
4. Ilokano	abút	hole, pit, perforation
5. Dumagat (Casiguran)	ʔəbút	hole; vagina
6. Aklanon	gabót	a hole (in anything except the ground)
7. Ifugaw (Batad)	ubut	to make a hole in thin material such as cloth
8. Yami	votovot	having a hole

*-but₃		husk, coarse hair or fiber (dbl. *-nut)
1. PWMP	*rambut	thread, fibers
2. PWMP	*sabut	coarse hair or fiber
3. Hanunóo	gábut	coconut husk
4. Sasak	jabut	hairy
5. Javanese	jəmbut	pubic hair
6. Karo Batak	mbutbut	fibrous, rough or stringy, as wood that has been cut with a dull knife

*-but₄		unclear, hazy, misty
1. PMP	*kabut	fog, haze, mist
2. Yakan	labut	cloudy (not clear, of vision)
3. Dusun (Kadazan)	lovut	dim, hazy, blur (of eyes)
4. Tiruray	rugabut	(of eyes) unable to see clearly because of irritation; (of the air) hazy

*-bux		dust
1. PMP	*debuh	dust
2. PMP	*labuh	ashen, grayish
3. PAN	*qabux	ashes
4. Arosi	agohu	dirt, dust (< *-puk₂?)
5. Amis	alafoh	dust blown by the wind
6. Manobo (Ata)	alibu	dust
7. Melanau (Balingian)	davəw	dust, ashes in hearth
8. Samoan	efuefu	dust, powder (< *-puk₂?)
9. Kambera	kaŋabu	dust, lumpy sand
10. Fijian	kuvu	dust, spray, smoke, steam (<*-puk₂?)
11. Malay	ləbu	dust
12. Wolio	ombu	smoke, steam, mist
13. Samoan	pefu	dust (<*-puk₂?)
14. Old Javanese	rabu	dust, dried mud
15. Bikol	talbó	dust

Note: The Oceanic members of this set (Arosi, Samoan, Fijian) may reflect *-puk₂ 'dust'.

The c-roots

*-cak₁		muddy; the sound of walking in sticky mud (dbl. *-Caq, *-tak, *-tek)
1. PWMP	*bacak	muddy, waterlogged (of ground)
2. PWMP	*bicak	muddy
3. PWMP	*lucak	to trample earth, turn into mud, as a paddy field prior to planting
4. PPH	*pisák	muddy
5. Toba Batak	birsak	squirting or spraying of water
6. Toba Batak	hasak	rushing, of water
7. Kambera	kapihaku	mud, muddy
8. Javanese	kracak	sounds of water pouring or gushing
9. Cebuano	lagasák	splattering, pattering noise made by water falling on something
10. Totoli	lampesak	mud
11. Malay	ləcak-ləcok	the sound of a man walking through sticky mud
12. Banjarese	licak	muddy
13. Karo Batak	oncak	to shake, slosh about, as water in a can that someone is carrying
14. Mansaka	pasak	earth, ground
15. Wolio	ranca	splash
16. Mandar	ressaʔ	mud
17. Sasak	ricak	marshy, swampy, muddy
18. Aklanon	támsak	to splash, splatter
19. Bikol	tapsák	a splash

Note: Sundanese *cakcak* 'gecko' may belong here, through the chirping sound it makes, which can resemble that of the sucking sound of sticky mud in walking.

*-cak₂		to stab
1. PAN	*cakcak	to hack, chop to pieces
2. Iban	pancak	to strike, stab, pierce
3. Yogad	sessak	to stab
4. Bikol	tagsák	to stab
5. Old Javanese	uncak	to stab, pierce

*-cek		blind
1. PWMP	*picek	blind in one eye
2. Kankanaey	budsák	blind, sightless

(continued)

3. Maranao	isək	blind, have poor eyesight
4. Ngaju Dayak	lusek	as if blind, asleep
5. Dumagat (Casiguran)	pəsə́k	be poked in the eye with a sharp-pointed instrument

Note: Also Javanese *picak* 'blind'.

*-ceq		to hatch, break into large pieces
1. PAN	*ceqceq	in pieces
2. PAN	*peceq	to break into several large pieces, as a hatched egg
3. Balinese	bəncah	to wreck, break in pieces; perish, piece, fragment
4. Mandar	bisse	to break open (as an egg or abscess)
5. Bikol	himsá?	to hatch
6. Paulohi	hisa	to split to pieces (glass, bamboo)
7. Paulohi	kosa	to shatter (egg)
8. Old Javanese	rəcah	to cut into pieces
9. Maranao	salasa?	to split, break
10. Maranao	sarisa?	to break, broken

*-cik		to fly out, splash, spatter (dbl. *-cit, *-cut)
1. PMP	*becik	to spatter, fly off in all directions
2. PWMP	*lagecik	to splash, splatter
3. PAN	*lecik	to fly off, of solid bits or water droplets
4. PWMP	*peRcik	to splash, spatter
5. PMP	*picik	to splash, spray, sprinkle
6. PWMP	*ragecik	to splash, spatter
7. PMP	*recik	speckled, stained
8. PWMP	*ricik	the sound of splashing, etc.
9. PPH	*talesik	to splash, spatter
10. PS	*təsik	to spurt, spray out
11. PPH	*wisik	sprinkling or spraying
12. Aklanon	ásik	to splash, spatter
13. Kankanaey	labsík	to overflow, flow over
14. Hanunóo	lásik	sudden flying off of splinters and the like
15. Amis	licik	small broken pieces that have flown and landed someplace
16. Singhi	masik	to sprinkle

(continued)

17. Ifugaw	pahík	splashing or spattering water
18. Kankanaey	palsík	hit, struck by a piece of wood, stone, etc., as when s.o. is cleaving wood
19. Keley-i	patsik	for small pieces of things to be thrown about due to the force of an action
20. Iban	pərəcik	spark
21. Tagalog	pilansík	to spatter, splash (of liquid)
22. Palawano	rebsik	splash
23. Sangir	sesi?	what spatters away or sprays out
24. Tagalog	tabsík	splash of liquid produced by s.t. rolling, as a wheel or a ball rolling upon a pool of water
25. Cebuano	talígsik	large raindrops which fall in a scattered way
26. Binukid	tampisik	to splatter, splath
27. Aklanon	támsik	to splash, splatter
28. Lau	taŋasi	to splash, sprinkle, wet with rain
29. Hanunóo	tarupásik	splash, splattering of a liquid, as water
30. Tiruray	tərəsik	(of liquids) to splash
31. Tagalog	tilabsík	to spatter (of hot liquid)
32. Tagalog	tilamsík	a splash of water; spark from a fire
33. Bikol	tipsík	a splash (smaller than tapsák)
34. Sambal (Botolan)	tohík	to sprinkle with water
35. Pangasinan	waisík	to sprinkle
36. Dumagat (Casiguran)	wasék	to flick water from the end of one's wet finger
37. Hanunóo	wísik	shaking, sprinkling

*-cit		to squirt out (dbl. *-cik, *-cut)
1. PMP	*becit	to squeeze, squirt out
2. PMP	*lecit	to squeeze out, squirt out
3. PWMP	*pa(n)cit	to splash, spray
4. PMP	*pe(R)cit	to squeeze, squirt out
5. PMP	*picit	to squeeze out
6. PMP	*puRcit	to spurt or gush out
7. Iban	ancit	to send flying, shoot out like a shot
8. Balinese	əncit	diarrhoea
9. Sundanese	gəncet	to press or squeeze out
10. Iban	kəcit	to spurt out, cause to spurt out
11. Malay	kincit	a slight escape of fecal matter
12. Bikol	kupsít	to spurt
13. Malay	lancit	to squirt out, of water in a thin stream

(continued)

14.	Ilokano	lúsit	to squeeze, press, as fruit or eggs
15.	Singhi	ŋisit	to squirt forth
16.	Malay	pəlancit	to shoot out, as when a seed pops out of its pod when the pod is squeezed
17.	Malay	paléncét	forced out by pressure
18.	Malay	pərancit	to fly about in all directions, as chips of stone struck by a bullet
19.	Kanowit	pəsit	to squeeze s.t. with the fingers so hat it will pop
20.	Ilokano	pugsít	to spurt out, gush out
21.	Bare'e	pulonci	glance off, slip off
22.	Bikol	pulsít	to slip away without being noticed; a spout
23.	Minangkabau	racit	to knead; to squeeze
24.	Ida'an Begak	rissit	spurting out
25.	Rejang	təkəcit, təkicit	involuntary defecation as a result of shock, sneezing or passing wind

Note: Also Iban *palasit* 'to slip out, shoot out, press out'.

*-cut			to squirt, squeeze or slip out (dbl. *-cik, *-cit)
1.	PWMP	*lecut	to slip, squirt or squeeze out; be born
2.	PWMP	*lucut	to squeeze or squirt out
3.	PS	*pəsut	to press out, squeeze out
4.	PMP	*pucut	to spurt out
5.	Ilokano	apsút	to slide out, slip, escape
6.	Itbayaten	arsot	being driven out by pressure, as in laying of eggs or delivery of child
7.	Cebuano	hípsut	to slip out
8.	Cebuano	ípsut	to slip out
9.	Malay	lancut	a gush of water; the sound of water spurting in large jets
10.	Keley-i	lubhut	to fall through (as a squirrel falling through the roof after jumping there)
11.	Malay	lucut	to become undone and slip away
12.	Ilokano	lugsót	to have the rectum protruding
13.	Kankanaey	lupsút	to slip out
14.	WBM	pansut	to squirt out
15.	Malay	pərancut	to fly about in all directions, as chips of stone struck by a bullet
16.	Ayta Abellen	polhot	to slip on, over, or into
17.	Bare'e	pulosu	slipped off, slipped down

(continued)

18. Malay	rucut	to slip off or slip down
19. Javanese	trucut	let something slip (in, out)
20. Bare'e	wisu	a forward-shooting movement; to shoot out (water, a bird flying suddenly from cover)

The C-roots

*-Caq		mud; earth, ground (dbl. *-cak.₁, *-tak, *-tek)
1. PMP	*buRtaq	earth, soil, mud
2. PPH	*lutáq	mud, ground, earth
3. PAN	*NiCaq	type of soil, clay
4. Itneg (Binongan)	píta	earth, ground
5. Amis	potaʔ	soil, soft mushy dirt
6. Amis	sotaʔ	earth, dirt; mud; land
7. Kambera	weta	mud, morass
8. Bungku	wita	earth, land

*-Ceg		to hit, beat, pound (dbl. *-Cug)
1. PAN	*CegCeg	to knock, pound, as a nail
2. PWMP	*keteg	beating of the heart; pulse
3. PWMP	*qe(n)teg	to beat against
4. Singhi	gatog	to tap
5. Dusun (Kadazan)	hontog	to shake (of floor); tap, knock, rap
6. Balinese	pantəg	to hit, beat, smack, bang
7. Lun Dayeh	teg	the sound of a fired shotgun, rapid heartbeat, etc.
8. Ilokano	teltég	to knock, thump, strike, beat
9. Dusun (Kadazan)	tontog	to tap, knock, beat, rap, shake

*-Cek		mottled pattern (dbl. *-Cik.₁)
1. PMP	*batek	mottled design (as of a tattoo)
2. PWMP	*pantek	spotted, dappled
3. PWMP	*putek	spotted
4. Ayta Abellen	bolitək	figured, spotted
5. Malay	burantak	speckled neck feathers (seen on ground doves and some other birds)

(continued)

6. Ibaloy	kaltek	having many small holes, as pock-marked skin; also of spots on the skin of a carabao or cow caused by insect bites
7. Ibaloy	maltek	spots, stains, speckles
8. Iban	pərantak	spotted
9. Tboli	tetek	freckles; (of fruit) brown spots on the skin before it rots

Note: Also PAN *paCak 'spotted, speckled, as the skin of an animal'.

*-Cem		dark, of color; a dark color (dbl. *-dem, *-lem)
1. PAN	*kuCem	cloudy, obscure
2. PWMP	*li(n)tem	deep black, shiny black
3. PMP	*qitem	black
4. Maranao	datəm	dark cloud; darkness
5. DPB	ətəm	dark blue, tint derived from indigo
6. Sarangani Bilaan	fitam	black
7. BM	kintom	dark brown
8. Agutaynen	lagtem	dirty and discolored, dark (as skin from too much sun exposure)
9. Kenyah	metem	dark
10. Maranao	rotəm	dark, as clouds in stormy skies

Note: Also Banggai *antoŋ* 'black clouds'.

*-Ceŋ		stinging nettle: *Laportea* spp.
1. PAN	*laCeŋ	stinging nettle: *Laportea* spp.
2. PMP	*lalateŋ	stinging nettle: *Laportea* spp.
3. PMP	*zalateŋ	stinging nettle: *Laportea* spp.
4. PMP	*zalateŋ	stinging nettle: *Laportea* spp.
5. Amis	lidatəŋ	Stinging Nettle tree: *Laportea pterostigma*

*-Cik.₁		mottled, spotted (dbl. *-Cek)
1. PWMP	*batik	make a design; tattoo
2. PAN	*beCik	tattoo
3. PPH	*butik	spotted, dappled, speckled
4. PMP	*getik	spotted, striped
5. PWMP	*patik	mottled pattern
6. PWMP	*pe(n)tik	spot, dot
7. PMP	*rintik	steady dripping of sweat or drizzling rain; dot, spot, speck
8. Tiruray	bərətik	freckle-like spots on any surface
9. Malay	bintek-bintek	small dark heat spots
10. Maranao	bortik	spot, dot
11. Makassarese	burinti?	spots or speckles on the skin
12. DPB	gotik	striped, spotted, of animals like the tiger
13. Tausug	lintik-an	covered with tiny patches or spots of two colors (esp. black and white, often of chickens)
14. Iban	pərintik	spotted, speckled
15. Pangasinan	potík	to decorate cloth, etc. with braided threads
16. Ilokano	ratík	spots of mold or mildew on wet tobacco leaves, damp cloth, etc.
17. Kankanaey	sampatik	with scattered speckles
18. Melanau (Mukah)	titik	spot, dot

Note: Also PAN *tiktik 'to tattoo'.

*-Cik.₂		to spring up; flicking motion
1. PWMP	*bala(n)tik	spring-set spear trap
2. PAN	*betik	to flick, fillip with finger
3. PMP	*bitik	snare, noose trap; to spring up suddenly
4. PWMP	*ketik	to snap the fingers
5. PWMP	*labetik	to flick or snap
6. PWMP	*letik	to flick with the finger
7. PWMP	*pale(n)tik	spring-set spear trap
8. PAN	*peCik	to snap, as the fingers, or a slingshot
9. PWMP	*pilantik	spring-set spear trap
10. PMP	*pitik	to fillip, flick with the finger
11. Binukid	baletik	to flick (s.t.) with the thumb and forefinger
12. Bontok	bolintik	a child's marble
13. Iban	gitik	shivering, quivering (as with cold)
14. DPB	gutik	to move momentarily, as a fishing line by the tug of a fish on the bait

(continued)

15. Cebuano	hutík	for a fish to snap at the bait
16. Iban	jəlǝntik	to fillip, flick with the fingernail
17. Ilokano	kallátik	to rebound; spring back on collision or impact
18. Mapun	kutik	to snap one's fingers (as to get the attention of s.o.)
19. Tiruray	latik	to snap one's fingers sharply against someone's shin
20. Sangir	linti?	to spring, spring loose
21. Ifugaw	pagtík	what is hurled away, what bounds away by itself as if it were exploding
22. Ifugaw	pátik	the act of flying away, of being flung away; applied only to splinters
23. Ifugaw	pultík	to fillip, snap with the finger
24. DPB	putik	to pick, as a fruit
25. Iban	sǝlǝntik	to flick
26. Dusun (Kadazan)	titik	to strike a match
27. Bikol	?ulaptík	to hop (as a grasshoppper); to jump from place to place (as insects)

*-Cug		to knock, pound, beat (dbl. *-Ceg, *-Cuk)
1. PAN	*CugCug	to knock, pound, beat
2. Ifugaw (Batad)	bitug	to bump oneself as by hitting the head against a beam, by falling
3. Ayta Abellen	hahtog	to fall hard on the ground
4. Bontok	kanítog	to make a solid, knocking sound, as the noise of two shields banging together
5. Bikol	karantóg	a thud; the knocking sound of loose mechanical parts
6. Balinese	kǝtug	a thud; the sound of thunder; to push, knock, bang the head
7. Bontok	kitóg	to produce a dull thud, as when an egg with a chicken inside is shaken
8. Hanunóo	líntug	banging s.t. (as the head) against s.t. else
9. Bikol	luntóg	a thud; to fall with a thud
10. Itbayaten	paltog	a gun, firearm
11. Sundanese	tegtog	to collide, as two objects
12. Aklanon	túetog	to pound, hit, knock
13. Javanese	ṭug	a knocking beat on a gamelan block
14. Sundanese	turugtug	a thundering sound; the sound of beating on a large drum
15. Tagalog	untóg	a bump on the head

*-Cuk		to knock, pound, beat (dbl. *-Ceg, *-Cug)
1. PAN	*balalaCuk	woodpecker
2. PMP	*butuk	to knock, pound, beat
3. PMP	*etuk	to knock, pound, thud
4. PMP	*ga(n)tuk	to knock against, of body parts
5. PWMP	*getuk	to knock, pound on
6. PWMP	*hagetuk	the sound of knocking
7. PMP	*katuk	to knock
8. PMP	*ketuk	to knock, pound
9. PWMP	*kutuk	to knock, pound
10. PWMP	*lagetuk	popping or knocking sound
11. PWMP	*latuk	to hit, pound, rap on s.t.
12. PWMP	*lutuk	the sound of breaking or pounding
13. PAN	*paleCuk	to shoot; sound of shooting or exploding
14. PAN	*patuk	to knock, strike against
15. PMP	*petuk	to beat, throb; pulse
16. PWMP	*pituk	to tap, rap
17. PMP	*rantuk	to hit with an instrument, bump into
18. PMP	*retuk	to hit, pound, chop
19. PPH	*suntuk	to punch
20. PWMP	*tuk	the sound of a knock
21. Uma	balintutu?	woodpecker
22. Sundanese	bəlatok	sound of an explosion, the crackling of fire, etc.
23. Sasak	dantuk	to kick against with the foot
24. Hawu	detu	to knock, beat
25. Ifugaw	duntúk	blow of one's fist
26. Ngaju Dayak	garutok	sound of rudder striking against boat
27. Malay	gəlatok	to chatter, of the teeth
28. Ilokano	goták	to palpitate, throb, pulsate violently
29. Balinese	intuk	to stamp, grind, pulverize
30. DPB	kastuk	to resound, of bamboo water containers that strike together when they are carried in fetching water
31. Old Javanese	kiṭuk	to make a knocking sound
32. Murut (Timugon)	kurutuk	to make a repeated clacking sound
33. Romblomanon	lagátuk	the clack of a shoe on a hard surface
34. Tagalog	lagutók	a short, sharp sound (as of end-bones—carpal or tarsal—snapping upon being stretched on hand or fingers)
35. Bikol	lapaták	clonk!; to make this sound
36. Sangir	lentu?	to explode, burst open with a bang, as chestnuts in hot ashes
37. Kankanaey	litók ~ litó?ok	to crack, snap (as finger joints)
38. Bikol	lugtók	to hit with s.t. clasped in the closed fist

(continued)

39. Bikol	luʔtók	to hit with s.t. clasped in a closed fist
40. Kambera	nútuku	to knock, beat, tap on
41. Old Javanese	palatuk	woodpecker
42. Bikol	putók	to blow up, burst, explode; to fire, as a gun
43. Javanese	rəkutu:k	a loud cracking sound
44. Ilokano	rittók	the crackling sound of the joints of the fingers
45. Makassarese	runtuʔ	the pounding of the heart
46. Malay	santok	knocking or tripping up against
47. Puyuma	saḷṭuk	to hit by chance
48. Iban	səntok	to collide, touch against
49. Ilokano	sintók	to knock on the head
50. Ayta Abellen	tadoktok	to knock or rap on a hard surface
51. Murut (Timugon)	talutuk	a bamboo instrument that is beaten like a gong
52. Tetun	tanutuk	to beat, to flog
53. Iban	tatok	to knock, tap, peck
54. Sangir	tətuʔ	to beat s.t. with a stone or a mallet, as tree bark
55. Tagbanwa (Central)	tuntok	to hit, box, punch with the fist
56. Ngadha	vitu	a rod for beating clothes, etc.; to beat with such a rod
57. Komodo	wetoʔ	woodpecker
58. Manggarai	wetuk	to hit with something

Note: Also PAN *tuktuk 'to knock, pound, beat'.

*-Cuŋ		deep resounding sound (dbl. *-taŋ, *-teŋ$_{-1}$, *-tiŋ)
1. PMP	*betuŋ	booming sound
2. PAN	*CuŋCuŋ	to make a booming noise
3. PWMP	*tuŋ	booming sound
4. BM	bolontuŋ	rumbling, roaring, or stamping sound
5. Malay	dəntoŋ	to boom; a deep echoing roar
6. Iban	əŋkəratoŋ	harp with four or more strings, consisting of a sounding box with highly decorated ends
7. Ngaju Dayak	garuntoŋ	thunder
8. Ida'an Begak	gətuŋ	fast beat (in playing gong to announce death)
9. Dusun (Kadazan)	ho-hotuŋ-on	gong
10. Cebuano	kalátuŋ	bamboo slit-gong
11. Kenyah (Long Wat)	katuŋ	drum

(continued)

12. Old Javanese	kəṇṭuŋ	to rattle, knock
13. Javanese	kluṇṭuŋ	metallic clanking
14. Malay	ləntaŋ-ləntoŋ	ding-dong
15. Kankanaey	litóʔoŋ	to bounce, bang, knock loudly
16. Makassarese	mattuŋ	to make a dull sound (as heavy fruit falling to earth), to rumble, roar (as distant cannon)
17. Manggarai	ntuŋ	k.o. frog that raises a loud chorus when it rains
18. Malagasy	róntona	fired at with many guns, fired at by a volley
19. Iban	səgəruntoŋ	to sound, rumble (e.g. stomach)

The d-roots

*-daŋ-1		to dazzle, shine (dbl. *-laŋ-1, *-naŋ, *-raŋ)
1. PPH	*sel(e)daŋ	to shine, be bright
2. PWMP	*sendaŋ	light, radiance
3. Malay	andaŋ	flare; torch of dry palm leaf
4. Toba Batak	dadaŋ	to shine, beam, radiate
5. Old Javanese	paḍaŋ	clearness, brightness, light
6. Maranao	siraŋ	to dazzle
7. Bikol	sudaŋ	day
8. Karo Batak	təndaŋ	coconut-leaf torch; to illuminate

*-daŋ-2		to warm by a fire
1. PAN	*daŋdaŋ	to warm by a fire
2. PAN	*tadaŋ	to expose to heat
3. Iban	daŋ	hot
4. Tae'	men-daraŋ	to warm oneself by a fire
5. WBM	hizazaŋ	to warm oneself at a fire
6. Lun Dayeh	idaŋ	the heat of the sun
7. Keley-i	ledaŋ	to heat metal in the process of forging (bolo, knife); to smelt metal
8. Saisiyat	papaṛaŋ	to roast over a fire

(continued)

9. Kavalan	szaŋ	sun
	Ri-szaŋ	to expose to the sun

Note: Tae' *men-daraŋ* cannot regularly reflect *daŋdaŋ, since the medial cluster is normally preserved where the first consonant is a nasal, as in PWMP *banban > *bamban* 'a tree, the bast fibers of which are used as binding material in house construction: Grewia laevigata', PMP *bunbun > *bumbun* 'to heap or bank up, fill up', or PWMP *buŋbuŋ 'bamboo tube used as a storage container' > *bumbuŋ* 'term of measurement used to specify the number of large bamboo containers that one takes of a liquid'.

*-daR		to lean on, recline (dbl. *-diR)
1. PWMP	*sandaR	to lean on
2. Ibaloy	aŋshal	to lean against s.t.; to rest one's back against s.t.
3. Bare'e	honda	a rack or railing on which s.t. is supported or leans
4. Maranao	indag, irag	to recline, tilt
5. Banggai	kandal	a railing, place to lean
6. Palauan	olsírs	to lean against
7. Bare'e	onda	rack or railing on which s.t. is supported or leans
8. Maranao	sindag, sirag	to lean against
9. Palauan	skors	cane, crutches, walking stick

Note: Also Puyuma *saŋɖal* < *saŋedaN 'to lean against'.

*-dek$_1$		hiccough, sob (dbl. *-gek)
1. PMP	*gedek	sound of a sob
2. PPH	*hendek	to moan, groan
3. PS	*Rinduk	to hiccup
4. PWMP	*si(n)dek	to sob, inhale suddenly
5. Kayan	herek	swallowing accidentally; gulping
6. Kayan	pidek	to sob
7. Bontok	ʔalindadək	hiccough
8. Kankanaey	saʔində́k	to sob
9. Dumagat (Casiguran)	ʔududə́k	to grunt, of a pig
10. Cebuano	yádʔuk	to swallow down a liquid

*-dek.₂		pulverized, pounded fine (dbl. *-bek.₂, *-dek, *-mek, *pek)
1. PMP	*dekdek	to pound; pulverized rice
2. PWMP	*ledek	to pound into powder, as grain
3. Kapampangan	daldák	to grind, pound, mill
4. Kapampangan	darák	rice bran
5. Sasak	ladək	fine (of meal, flour)
6. Sasak	padək	to beat with a stick
7. Tiruray	ʔədək	to pound, as rice with a pestle

*-dem.₁		dark; overcast (dbl. *-Cem, *-lem)
1. PAN	*demdem	gloom, darkness; dark, overcast, gloomy
2. PAN	*dudem	obscured by clouds
3. PWMP	*edem	overcast; dull lustre
4. PMP	*gu(n)dem	overcast, darkened
5. PPH	*ke(n)dem	dark; darkness
6. PMP	*kidem	to close the eyes
7. PAN	*kudem	darkened by clouds, overcast
8. PMP	*le(n)dem	shaded, shadowy
9. PWMP	*lidem	dark in color; dim
10. PWMP	*ludem	dim, overcast
11. PWMP	*medem	dark, obscure
12. PWMP	*padem	to exinguish, douse a fire
13. PWMP	*pidem	to close the eyes
14. PWMP	*qi(n)dem	dark, black
15. PAN	*qudem	dark, dull (of light or color)
16. PWMP	*sedem	dark
17. PWMP	*sidem	obscure, dusky
18. PAN	*tedem	dark, darkness
19. PMP	*tidem	dark, obscure; black
20. PPH	*tudem	dark spot on body
21. Tausug	andum	cloud (dark)
22. Javanese	burəm	(of the moon) hazy, clouded over
23. Ayta Abellen	dəgləm	evening; become nighttime
24. Javanese	jarəm	bruised, black and blue
25. Ilokano	kiddém	to tarnish, become dull
26. Hawaiian	kilo	first night of the new moon
27. Melanau (Mukah)	memadem	overcast
28. Palauan	miremérem	dark (at night)
29. Bare'e	mudo	overcast, clouded over
30. Pangasinan	ŋárəm	afternoon
31. Sangir	peduŋ	dark (of wrath)
32. Puyuma	ʔudədəm	black
33. Iban	rədam	dark red, maroon
34. Ilokano	na-ridém	dim, overcast

(continued)

35. Bikol	rumárom	cloudy, overcast; leaden (the color of the sky on stormy days)
36. Masbatenyo	rumiróm	gloomy, forboding
37. WBM	rusirəm	dark
38. Bikol	sinárom	twilight; dusk
39. Balinese	surəm	obscure, dim
40. Puyuma	uDeDem	black

Note: Also Casiguran Dumagat *lədúm* 'shade; shady', Kapampangan *dalúmdum* 'darkness'.

*-dem.₂		to brood; begrudge; remember; keep still
1. PAN	*demdem	to think, ponder, consider
2. PMP	*handem	to think, understand
3. PWMP	*hidem	to keep s.t. to oneself; silent, secret
4. PWMP	*qendem	to think, ponder; brood
5. PMP	*se(n)dem	to brood, think, be silent
6. PWMP	*ta(n)dem	to remember
7. PS	*taRəndum	memory; to remember, think of
8. PWMP	*tendem	to recall, recollect, remember
9. Javanese	aŋrəm	to sit brooding
10. Bare'e	endo	reminiscing upon what is gone
11. Iban	gərindam	thoughts
12. Bikol	girumdóm	to reminisce about, to call to mind; to commemorate
13. Tboli	hendem	think, ponder
14. Maranao	ləmdəm	hatch – hen or bird
15. Toba Batak	podom	to brood, sit on eggs
16. Palawano	rendem	memory, recollection
17. Javanese	siḍəm	quiet, silent, withdrawn
18. Lau	tago	to think

Note: Also PAN *nemnem 'to think', Kayan *selem* 'to think within oneself'.

*-deŋ		straight; to straighten (dbl. *-teŋ.₃)
1. Keley-i	ʔagdəŋ	straight
2. Ifugaw (Batad)	andoŋ	for s.t. crooked to become straight, as a tree, person's posture, etc.
3. Iban	kədaŋ	straight; to straighten (esp. wood by use of fire)
4. Subanon	mo-logdoŋ	straight
5. Batak (Palawan)	ma-taldɨŋ	straight
6. Tboli	tədəŋ	to make straight; to justify
7. Molbog	mo-tiddoŋ	straight

*-det		packed in, compressed
1. PAN	*detdet	packed tightly together
2. PWMP	*pa(n)det	compact, compressed; packed in tightly
3. PMP	*pedet	to pack in tightly
4. PAN	*qedet	to press down on
5. PWMP	*sandet	packed tightly
6. PWMP	*sendet	tightness, congestion
7. Bikol	budát (+M)	full and unable to eat any more
8. Bikol	gadót	compact, compressed
9. Dumagat (Casiguran)	kədə́t	thick underbrush
10. Manggarai	piret	too tight (of clothes)
11. Bontok	ʔidát	to pack mud on dykes of pondfield
12. Ilokano	sidét	dense, thick, close, compact, crowded (of plants, writing, etc.)
13. Sundanese	tɨndɨt	to ram down, batter, pound (as the ground, or as to reduce unevenness)

*-dik		small (dbl. *-tik$_{-2}$)
1. PMP	*kedik	small in size or amount
2. Balinese	bədik	little, not much, a bit
3. Dusun Deyah	idik	small
4. Dusun Malang	kadik	small
5. Old Javanese	kidik	few
6. Manggarai	merik	small
7. Dusun Witu	odik	small

Note: Also Totoli *dedek*, Dampelas *idek* 'small'.

*-diR		to lean against (dbl. *-daR)
1. PMP	*sandiR	to lean against, lean or incline on
2. PMP	*sendiR	to lean on or against
3. PWMP	*sindiR	to lean on or against
4. Ibaloy	saŋshil	to lean back

*-du		ladle, spoon (dbl. *-duk)
1. PAN	*Cidu	spoon, ladle, dipper
2. PAN	*sidu	spoon, ladle
3. PMP	*sudu	spoon, ladle, scoop
4. Tagbanwa (Central)	karo	ladle

*-duk		ladle, spoon (dbl. *-du)
1. PWMP	*sanduk	spoon, ladle
2. PWMP	*senduk	spoon
3. PMP	*si(n)duk	spoon, ladle
4. PMP	*suduk	spoon, ladle
5. Malay	cedok	spooning up; digging up with a spade
6. Erai	hahuru	spoon
7. Buli	iru	coconut shell ladle
8. Tetun	kanedok	ladle, spoon
9. Kambera	kinjiru	spoon used for mixing
10. Dumagat (Casiguran)	saldúk	to dip or scoop up
11. Penan (Long Labid)	tarok	spoon

Note 1: Also Kayan (Uma Juman) *huluk* 'spoon, ladle'.
Note 2: Source for Penan (Long Labid) *tarok* unknown.

*-duŋ		to shelter from rain or sun; head cover (dbl. *-luŋ-₂)
1. PWMP	*ci(n)duŋ	shelter
2. PMP	*duŋduŋ	sheltered, as from wind, rain or sun
3. PWMP	*kanduŋ	womb
4. PWMP	*kuduŋ	head cover for women
5. PWMP	*leduŋ	to shelter, protect
6. PAN	*liduŋ	shelter, cover, protection; shade
7. PPH	*pandúŋ	head covering
8. PMP	*ru(n)duŋ	to shelter
9. PWMP	*siduŋ	sheltered place
10. PPH	*taduŋ	sun hat, used to provide shade
11. PMP	*teduŋ	to take shelter, cover the head
12. PMP	*tiduŋ	to cover, shelter, protect
13. PMP	*tuduŋ	head cover; protection from sun or rain
14. PMP	*uduŋ	hut, temporary shelter
15. Keley-i	aliduŋ	to give shade (said to be 'archaic')
16. Toba Batak	bahuduŋ	field hut
17. Tagbanwa (Central)	baloŋ	a small temporary shelter in one's fields
18. WBM	buriruŋ	a shade for a lamp, esp. when hunting at night to keep the lamp from blinding the eyes
19. Tae'	buruŋ	rain cape of bamboo or pandanus leaves, worn on the back with a cowl covering the head
20. Sangir	galinduŋ	to protect, screen, shelter
21. Ifugaw	halídoŋ	large circular hat for wear in the rain
22. Malagasy	híndona	take shelter from the rain or sun

(continued)

23. Kankanaey	íduŋ	to shelter
24. Javanese	kroḍoŋ	protective covering
25. Kavalan	razuŋ	shade; to take shelter from the rain
26. Kayan	lemiruŋ	to shelter or shield s.t. from wind
27. Ngadha	ludu ~ ludhu	to cover, screen, protect, veil
28. Tae'	mentireruŋ	to shelter
29. Chamorro	nuhoŋ	shady, shaded, shadowy
30. Maranao	podoŋ	native hat; top of hat
31. Amis	raroŋ	a hut, a temporary shelter, as when it is raining
32. Tiruray	rinduŋ	to take cover temporarily
33. Malagasy	salóndona	the act of covering the head
34. Malay	səlindoŋ	veiling oneself; hiding behind a veil of any sort
35. WBM	sərəruŋ	to make a temporary shelter
36. Iban	sərudoŋ	a cover, shelter, temporary hut
37. Sasak	sunduŋ	a cloth or leaf covering wrapped around (growing) fruits to protect them from birds and bats
38. Gorontalo	woluŋo	shelter

*-duR		thunder; the rumbling of thunder (dbl. *-gur, *-kur, *-ŋur)
1. PWMP	*duRduR	thunder
2. PMP	*enduR	thunder
3. Aklanon	daeugdug	the rumbling of thunder
4. Bare'e	gundu	thunder
5. Bontok	kidól	thunder; to rumble, of thunder
6. Bikol	ʔuróg-ʔudóg	a thundering or rumbling sound

The g-roots

*-gak$_{-1}$		proud; to boast
1. PWMP	*ba(ŋ)gak	foolish
2. Malay	begak	foppish, dressed up
3. Aklanon	búgʔak	proud, haughty, boastful
4. Aklanon	hágak	bragging, boastful, proud
5. Balinese	laŋgak	proud, vain, conceited
6. Minangkabau	paŋgak	to pride oneself on s.t.; swagger

(continued)

7. Malay	segak	of women, dressing up; of a man, showing off his strength
8. Manggarai	sehak	proud, haughty, arrogant
9. Minangkabau	soŋak	handsome, showy
10. Tagalog	ʔugák	foolish, stupid

*-gak₂		raucous, throaty sound (dbl. *-gek, *-gik, *-guk)
1. PMP	*gakgak	to caw (crows), cackle
2. PWMP	*hagak	raucous throaty sound
3. PWMP	*lagak	sound of gurgling water
4. Mapun	aggak ~ ggak	a belch or burp
5. Bikol	bulák-bugák	to gurgle, gush out
6. Ngaju Dayak	gak	to thud (as of a fist striking an object, something falling)
7. Sundanese	gokgak	in a blatant tone, as in scolding s.o.
8. Kankanaey	ígak	break out into laughter
9. Mansaka	logak	to squeal (as a pig)
10. Agutaynen	ogak	crow, raven
11. Iban	paŋgak	frog
12. Maranao	raŋgak	sound made by difficult breathing, as in asthma, or when dying
13. Bikol	rigák-rigák	the sound of a croaking frog
14. Kapampangan	sagakgak	to guffaw
15. Sundanese	səŋgak	to shout, exult
16. Toba Batak	sigak	crow (si gak?)
17. Kankanaey	tagák	to cackle
18. Kapampangan	tugák	frog
19. Kankanaey	uágak	to low, moo

*-gaŋ		dry
1. PWMP	*egaŋ	dry, dried up
2. PWMP	*gaŋgaŋ	to heat or dry s.t. near a fire
3. PWMP	*paŋgaŋ	to roast, dry fry
4. PWMP	*tegaŋ	dry, dehydrated
5. Iban	agaŋ	dry branches and twigs of fallen tree
6. Sika	begaŋ	burned, very dry
7. Bikol	ʔigáŋ	referring to dried salt
8. Tiruray	lagaŋ	to dry over a fire

(continued)

9. Iban	ləgaŋ	to set, harden, as cooked rice when the water has steamed off; dry, as meat drained of blood
10. Kankanaey	lokgán	dry; empty; drained
11. Bukat	magaŋ	dry
12. BM	rogaŋ ~ yogaŋ	thirst
13. Bikol	tigán	almost dry (as a stream with very low water, or only patches of water)

*-gaw		confused, disoriented, lost (dbl. *-ŋaw$_{-1}$, *-taw$_{-2}$)
1. PWMP	*begaw	confused, disoriented by loud noise
2. PWMP	*regaw	to wander, stray
3. PWMP	*sigaw	disturbance, confusion
4. Bikol	lawgáw	confused, disoriented, off course, lost
5. Maranao	ligaw	lose one's way, confused in direction
6. Iban	səgau-səgau	wandering

*-gek		dull throaty sound (dbl. *-dek$_{-1}$, *-gak, *-gik, *-guk)
1. PWMP	*gek	the sound of a thud
2. PMP	*gekgek	animal sound
3. PWMP	*se(ŋ)gek	to choke when eating or drinking
4. PWMP	*sigek	to choke, sob
5. Ifugaw	boŋgók	to snore while sleeping
6. Kankanaey	buŋgók	to snore
7. Kankanaey	daŋgák	hoarse; to sound harshly, as the throat when one coughs
8. DPB	dərgək	to gulp, swallow
9. Kankanaey	duŋgók	to snore
10. Ifugaw	igók	gurgling sound produced by someone who drinks water
11. Kayan	kigek	noisy laughter
12. Iban	laŋak	to gulp down, swallow noisily
13. Sasak	ləgək	a gulp (of water); to swallow
14. Binukid	ligek	(for a child) to gasp, struggle for breath
15. Manggarai	reŋgek	to drone (of speech, etc.)
16. Agta (Eastern)	sagek	to gag, choke on s.t.
17. Dusun (Kadazan)	togok	to cluck (as a hen)

*-gem		to grasp in the fist (dbl. *-kem.₁)
1. PWMP	*agem	to hold, grip
2. PMP	*eŋgem	to hold s.t. in the mouth
3. PMP	*gem	to hold in the fist
4. PAN	*gemgem	fist; to hold in the fist
5. PWMP	*kugem	to hold in the fist
6. PWMP	*Ragem	clenched in the fist
7. Sundanese	caŋgɨm	a handful, fistful
8. Bare'e	gaŋgo	to grasp with the hand, as when seeking fish in the mud
9. Bikol	gugóm	a clenched fist
10. Kayan (Uma Juman)	higəm	to hold; carry in the hand
11. Ilokano	iggém	to hold; have or keep in the grasp
12. Kenyah	keregem	to clench the fist
13. Melanau (Mukah)	ñagem	fist, hand
14. Cebuano	sákgum	to hold something in both hands
15. Kenyah (Long Anap)	seŋgem	to grip, hold tightly
16. Javanese	taŋgəm	steel vise

Note: Also Banggai *gaŋgum* 'fist', Manggarai *regum* 'to hold in the fist'.

*-geŋ		to hum, buzz (dbl. *-guŋ)
1. PWMP	*eŋgeŋ	to moan, hum
2. PMP	*geŋ	a dull resounding sound
3. PMP	*re(ŋ)geŋ	dull resonance
4. Agutaynen	begeŋ-begeŋ	the indistinct sound of people talking
5. Manggarai	ceŋgeŋ	to buzz (of a bee)
6. Binukid	lageŋ	sound, voice, tune
7. Maranao	rigəŋ	to roar
8. Maranao	sigəŋ	peal, snore, vibration
9. WBM	yəgəŋ	to hum a tune

*-ger		to shake, shiver, tremble (dbl. *-ter, *-tiR)
1. PAN	*gerger	to shake, shiver, tremble
2. Sika	beger	to move; shudder, tremble, shake
3. DPB	dəgər	to shake (as a tree, to make its fruit fall)
4. Kayan	gen	shaking or vibration
5. Sasak	laŋgər	to shiver from the cold
6. Kayan	tegen	to shake, shaking; to vibrate

Note: Also PS *kəkiR 'to shiver, tremble'.

*-get	angry (dbl. *-git, *-gut, *-ŋaC, *-ŋeC, *-ŋiC, *-ŋuC)	
1. PWMP	*li(ŋ)get	to gnash the teeth in anger or impatience
2. Old Javanese	gəgət	to grit the teeth (with fury)?
3. Manggarai	jeget	angry
4. Hiligaynon	págut	to rave, be furious, gnash the teeth while asleep
5. Manggarai	riget	have an unpleasnt facial expression
6. Ilokano	supagét	to show or manifest anger (said of persons and animals)
7. Aklanon	ugót	be angry; get peeved
8. Manggarai (West)	weget	spiteful, malicious, peevish; hateful

Note: Also Balinese cəkət-an 'sensitive, short-tempered'.

*-gik	shrill throaty sound (dbl. *-gak, *-gek, *-guk)	
1. PMP	*egik	high-pitched animal sound
2. PMP	*gikgik	to titter, giggle
3. PPH	*úgik	to squeal, as a pig
4. Cebuano	agíʔik	to creak, squeak
5. Ifugaw	dagík	noise
6. Kankanaey	galokígik	to neigh
7. Cebuano	ígik	for pigs to squeal
8. Malay	rəgik	a gulp, small mouthful
9. Dumagat (Casiguran)	səgék	to gag (swallow s.t., choke on it, and cough it back up again)
10. Hiligaynon	tígʔik	shriek, choking sound

*-gis	to scrape (dbl. *-kis.$_1$)	
1. PAN	*gisgis	to rub, scrape against
2. Tausug	kagis	to scrape s.t. with a knife or similar object in order to clean it
3. Aklanon	lísgis	scratch, surface cut, abrasion
4. Aklanon	úgis	scraped clean of hair

*-git	angry, anger (dbl. *-get, *-gut, *-ŋaC, *-ŋeC, *-ŋiC, *-ŋuC)	
1. PPH	*iŋit	envy, envious
2. Kapampangan	giligít	trembling internally, tense with anger
3. Kankanaey	gitgít	to resent, be offended, get angry
4. Malagasy	móhitra	to dispute angrily
5. Kankanaey	siŋít	to tease, plague; trouble, vex

*-guC		to pull with a jerk (dbl. *-buC, *-NuC, *-zut)
1. PPH	*bagut	to pull out, as hair
2. PAN	*guCguC	to pluck, pull out
3. PWMP	*pagut	to pull or yank, as s.o.'s hair or tubers from the ground
4. PWMP	*raŋgut	to snatch at, pull with a jerk
5. Malay	cagut	to peck (of a fowl), pull (of fish at a bait)
6. Mansaka	daŋgot	to pluck; to pick (as a leaf)
7. Kapampangan	gígut	to chew, break seeds open with teeth
8. Ngaju Dayak	haŋgut	what is pull out (grass, post, person from water, etc.)
9. Tausug	hugut	to pull (nits from the hair)
10. Sasak	pugut ~ puŋgut	to pull the hair
11. Malay	rəŋgut	snatching at, tugging at; tearing at

*-guk		deep throaty sound (dbl. *-gak, *-gek, *-gik)
1. PWMP	*ceguk	hiccough
2. PMP	*de(ŋ)guk	deep throaty sound
3. PMP	*e(ŋ)guk	to make a gurgling sound
4. PMP	*gukguk	animal sound
5. PWMP	*haguk	deep throaty sound
6. PMP	*heqeguk	deep guttural sound
7. PPH	*iŋguk	throaty sound
8. PWMP	*leq(e)guk	to gulp, swallow
9. PMP	*re(ŋ)guk	deep throaty sound
10. PWMP	*seguk	to make a gurgling sound in the throat; hiccough
11. PPH	*sigúk	to choke, sob
12. PMP	*teguk	to gulp, swallow all at once
13. DPB	bəlguk	to swallow
14. Cebuano	búguk	to take a mouthful of liquid to gargle
15. Ifugaw	daŋgók	to snore while sleeping
16. Ibaloy	digok	esophagus
17. Lun Dayeh	gok	the sound made by a pig or boar
18. Kambera	haŋguku	individual movement of swallowing
19. Hiligaynon	huragúk	to snore
20. Bikol	laʔgók	to gulp down liquids; to wolf down food
21. Ifugaw	paŋgúk	stifled, suffocated because of swallowing s.t. the wrong way
22. Bontok	ʔosgók	to grunt, of a pig
23. Iban	ragok	hoarse; to speak hoarsely
24. Manggarai	ruguk	continuous clatter
25. Old Javanese	təlaguk	making the sound of gulping or swallowing

*-guŋ		deep resounding sound (dbl. *-geŋ)
1. PWMP	*aguŋ	gong
2. PWMP	*eguŋ	gong
3. PAN	*guŋguŋ	deep resounding sound
4. PWMP	*leguŋ	booming sound
5. PMP	ru(ŋ)guŋ	thunder
6. PAN	*teguŋ	to boom, resound
7. Ngadha	bhegu	to beat, drum, call to the dance
8. Iban	dəgoŋ	to reverberate, roar (as the voice of s.o. shouting in a cave)
9. Bikol	gayuŋgóŋ	a rattling sound
10. Bontok	goŋógoŋ	to reverberate, of the sound of a large gong
11. Bikol	hukrágoŋ	a snore
12. Cebuano	kagún	to make a hollow sound when empty
13. Bontok	kalígoŋ	to make a hollow sound, as when an empty can is dropped; to clank
14. Tiruray	kəraguŋ	thunder, thundering
15. Kankanaey	kígoʔoŋ	to rattle, clash, bang, bounce
16. Kankanaey	kíguŋ	to clink, ring, clang, clank
17. Tboli	loguŋ	deep rumble of thunder
18. Pangasinan	ógoŋ	to thunder (of any noise)
19. Old Javanese	reguŋ	to make a loud noise, trumpet, bellow (as a furious elephant)
20. Dumagat (Casiguran)	tagúŋ	to howl, of a dog
21. Ifugaw	tikgóŋ	bumping sound produced when s.t. that is hard falls and collides

*-gur		to purr; rumble (dbl. *-duR, *-kur, *-ŋur)
1. PMP	*degur	droning or humming sound
2. PMP	*gur	purring or grunting sound
3. PAN	*gurgur	sound of boiling water, splashing rain, etc.
4. PMP	*le(ŋ)gur	thunder
5. Sika	bagur	to beat a drum
6. Soboyo	baruŋgu	the rumble of thunder
7. Balinese	cəgur	to make the sound 'gur' (a gong hit with a hammer)
8. Erai	gogu	thunder
9. Paiwan	gurugur	a dog's ordinary bark
10. Sasak	iŋgur	to rumble, roar
11. Sundanese	jəgur	sound of thunder, cannon
12. Toba Batak	roŋgur	thunder
13. Malay	səŋgor-səŋgor	the growl or grunt of an animal

(continued)

14. Bintulu	siŋgur	to snore
15. Malay	tagur	a deep muffled sound like distant thunder
16. Bintulu	təgur	thunder

*-gut.₁		angry, annoyed (dbl. *-get, *-git, *-ŋaC, *-ŋeC, *-ŋiC, *-ŋuC)
1. Tagalog	bagót	fed up, impatient
2. Isneg	kéxut	anger
3. Manggarai	regut	angry (face)
4. Malay	ugut	browbeating; bluffing; threatening

*-gut.₂		to gnaw
1. PWMP	*gigut	to bite
2. PAN	*gutgut	front teeth, incisors
3. PWMP	*pagut	to snap at with the mouth
4. PWMP	*regut	to bite, nibble
5. Balinese	a(ŋ)gut	to snatch at s.t., bite at (dog)
6. Balinese	cəgut	to bite, gnaw
7. Cebuano	ígut	to scrape s.t. by rubbing a knife up and down against it
8. Balinese	puŋgut	to bite, bite off, put one's teeth into
9. Balinese	səgut	to bite, nibble (men and animals)

The h-roots

*-hak		clearing of throat (dbl. *-kak)
1. PWMP	*bahak	to laugh loudly
2. PWMP	*hakhak	to laugh loudly
3. Hanunóo	akhák	expectoration of cattarhal phlegm
4. Cebuano	údhak	deep cough or gasp for breath of the sort one gets from an asthma attack

*-huR		to mix, as food with water (dbl. *-bur, *-buR, *pur)
1. PWMP	*bahuR	to mix foods, as in preparing pig's fodder
2. PPH	*sahuR	to mix; a mixture

(continued)

3. Cebuano	áhug	to mix s.t. wet with s.t. dry
4. BM	kaug	to stir, mix
5. Palawano	lehug	to mix, mixed

The k-roots

*-ka		to split, force open (dbl. *-kaq.₂)
1. PMP	*beka	to split, crack open
2. PWMP	*bika	potsherd
3. PMP	*peka	to separate, disconnect
4. Bikol	raŋká	to force or pry open
5. Agutaynen	rika	a crack in a piece of wood, the ground, or in the skin or lips; a small crevice
6. Karo Batak	taka	to split, cleave
7. Itbayaten	vaka	cracking or cutting open
8. Paiwan	velaka	a split in wood

*-kab.₁		to open, uncover
1. PWMP	*e(ŋ)kab	to open, uncover
2. PWMP	*hu(ŋ)kab	to open, uncover
3. PWMP	*iŋkab	to open, uncover
4. PAN	*lekab	to open, uncover
5. PWMP	*li(ŋ)kab	to open, uncover
6. PWMP	*lukab	to turn up at the edge, expose the bottom side
7. PWMP	*ruŋkab	to open, peel off
8. PMP	*siŋkab	to open, uncover
9. PMP	*su(ŋ)kab	to open, break open
10. PPH	*tiŋkáb	to pry open
11. PWMP	*tu(ŋ)kab	to break open, force open
12. Bare'e	boka	to open by loosening the binding
13. Dusun (Kadazan)	haŋkab	to come off, come loose (as skin, s.t. pasted on)
14. Tagalog	hikáb	a yawn
15. Binukid	saŋkab	for a fish to open its mouth to catch food
16. Wolio	soka	to unfasten, undo, untie
17. Keley-i	təkab	to peel things apart that are stuck together
18. Cebuano	ukáb	to open with an upward or lifting motion

Note: Also Kadazan Dusun *hiŋkib* 'open slightly', Aklanon *tukíb* 'open up a little (as a window)'. Tagalog *hikáb* next to *higáb* 'a yawn' may show contamination from *-kab.

*-kab₋₂		to snap at with the teeth
1. 5. PPH	*sa(ŋ)kab	to seize prey in the water, of a crocodile
2. Hiligaynon	ʔáŋkab	of animals, to snap or bite, to seize
3. Hanunóo	hakáb	devouring, gobbling up, as food by
4. Pangasinan	kabkáb	to gnaw
5. Ilokano	rikáb	to gasp, respire (convulsively)
6. Bikol	tákab	to snap at (of animals), to nip

*-kad		to prop, support (dbl. *-ked, *-kud)
1. PWMP	*sekad	to prop, support
2. PWMP	*tu(ŋ)kad	prop, support; staff
3. Kambera	hiŋga	prop, support
4. Hawu	njuka	prop, support
5. Rotinese	saka	to prop up with something
6. Ilokano	saroŋkád	to prop, to plant the feet firmly against the ground

*-kak		to cackle, laugh loudly (dbl. *-kek, *-kik, *-kuk₋₂)
1. PMP	*akak	cackling laughter
2. PWMP	*cekak	to choke
3. PWMP	*dekak	shriek, as of laughter
4. PMP	*kak	sound of a cackle, loud laughter, etc.
5. PAN	*kakak	to caw, cackle
6. PWMP	*lekak	to cackle with laughter
7. PWMP	*pekak	to cackle, as a hen that has laid an egg
8. PMP	*pikak	to make a harsh throaty sound
9. PMP	*qekak	to choke, gasp, struggle for breath
10. PWMP	*sukak	a harsh throaty sound
11. PWMP	*tekak	to cackle
12. PPH	*tukák	kind of loud frog
13. Bikol	ʔagakʔák	clucking sound; to cluck
14. Javanese	bəkak-bəki:k	to shriek repeatedly
15. Hawu	hika	morning bird, named from its cry
16. Malay	iŋkak	to chuckle, guffaw
17. Ayta Abellen	kokak	croaking of frogs
18. Manggarai	ŋkehak	to gag of one about to vomit
19. Manggarai	ŋkéhak	to cough repeatedly
20. Kankanaey	pakák	to cackle
21. Amis	ʔanikak	crow
22. Lamaholot	rəkak	to shout, scream
23. Amis	səkak	the crow of a rooster; to crow

(continued)

24. Tagbanwa (Kalamian)	talaŋkak	to laugh
25. Kambera	taŋgaka	to roar with laughter
26. Agutaynen	telkak	to gag and cough when eating s.t. that is very acrid or strong tasting

*-kaŋ.₁		to bark, croak, hoot (dbl. *-keŋ.₂, *-kiŋ, *-kuŋ.₂)
1. PWMP	*ekaŋ	to hoot, croak
2. PMP	*kaŋ	animal sound
3. PAN	*kaŋkaŋ	resounding sound
4. Malay	dəŋkaŋ	croaking sound such as the note of the bullfrog
5. Malay	doŋkaŋ	a frog, sp. unident.
6. Kankanaey	kúkaŋ	to cry out, scream, screech

*-kaŋ.₂		to spread apart, as the legs
1. PWMP	*akaŋ	to take a long step, step over s.t.
2. PAN	*bakaŋ	bowlegged
3. PMP	*be(ŋ)kaŋ	to spread apart, as legs, or unbent fishhook
4. PWMP	*bikaŋkaŋ	to spread apart, as legs or an unbent fishhook
5. PWMP	*bi(ŋ)kaŋ	to spread apart, as the legs
6. PWMP	*bukaŋkaŋ	to spread apart, as the legs or an unbent fishhook
7. PWMP	*eŋkaŋ	to spread the legs
8. PWMP	*iŋkaŋ	to walk with legs wide apart
9. PWMP	*kaŋkaŋ	to spread the legs wide
10. PWMP	*la(ŋ)kaŋ	to spread the legs
11. PMP	*le(ŋ)kaŋ	to separate, disunite
12. PWMP	*pakaŋ	wide apart, of the legs
13. PMP	*pekaŋ	to stretch open or apart
14. PMP	*pikaŋ	to spread wide apart, as the legs
15. PWMP	*pukaŋ	crotch
16. PWMP	*raŋkaŋ	forked, as a branch
17. PWMP	*rekaŋ	wide apart (as the legs)
18. PWMP	*Ra(ŋ)kaŋ	to spread the legs
19. PMP	*sakaŋ	bowlegged
20. PPH	*saká?aŋ	to walk with legs wide apart
21. PMP	*sekaŋ	crossbar, strut, s.t. to hold things apart
22. PWMP	*sikaŋ	wide apart, of the legs
23. PPH	*tákaŋ	to spread wide open
24. PWMP	*tekaŋ	a step, a stride

(continued)

25. PWMP	*ti(ŋ)kaŋ		spread apart, as the legs
26. PWMP	*u(ŋ)kaŋ		spread open, as the legs
27. PMP	*waŋkaŋ		to spread the legs apart
28. PWMP	*zaŋkaŋ		wide apart, as the legs
29. Hanunóo	balakáŋ		space between one's legs
30. WBM	bivikaŋ		to squat with knees wide apart
31. Malay	boŋkaŋ		prone; astretch (in contrast to lying huddled up)
32. Sasak	cakaŋ		with fingers spread open
33. Balinese	cəŋkaŋ		span (tip of the first finger to thumb)
34. Rejang	cikaŋ		to stand with legs apart
35. Malay	cukaŋ		to split bamboos against a wedge
36. Iban	dədaŋkaŋ		to squat or kneel with hands on knees
37. Madurese	dharakkaŋ		sit with legs wide apart
38. Balinese	duŋkaŋ		step, stride, step over, take a big step
39. Sambal (Botolan)	halakáŋ		to split and spread apart the end of bamboo and make a hen's nest of it
40. Malay	jaŋkaŋ		generic for trees with stilt roots
41. Malay	jeŋkaŋ		extension of a leg; as when a mother draws out a child's leg to clean it
42. Bikol	kaʔáŋ		to walk with the legs spread apart
43. Bikol	karaŋkáŋ		describing the open hand with fingers spread apart
44. Malay	kəlaŋkaŋ		perineum; junction of lower limbs
45. Iban	liŋkaŋ		to walk or dance as if drunk, stagger, totter
46. Ida'an Begak	pədakkaŋ		to fall backward when sitting
47. Kelabit	pekakaŋ		spread apart, stretched (of forceps that have been forced apart)
48. Iban	pərəkaŋ		wide apart, wide open; sit sprawling
49. Javanəse	pleŋkaŋ		to inadvertently do a split
50. Iban	rikaŋ		swaggering; walk with feet apart
51. Tiruray	səbəkaŋ		v-shaped
52. Dumagat (Casiguran)	ségkaŋ		bow-legged
53. Toba Batak	sehaŋ		spread apart, of the legs
54. Balinese	səlaŋkaŋ		stand astride of
55. Kayan	səpakaŋ		to place a child or hold s.t. between one's legs
56. Iban	siŋkaŋ		step, pace
57. Kenyah	tapikaŋ		a v-shaped piece of wood used in constructing the platform on which a corpse is laid out
58. Balinese	tuŋkaŋ		to stride, straddle

Note: Next to *jaŋkaŋ* 'generic for a trees with stilt roots', Malay has *tər-jaŋkaŋ* 'wide astretch, of the limbs'. Also Balinese *əŋgaŋ* 'stand apart like a pair of compasses, legs'.

*-kaŋ₋₃		stiff, rigid; cramps
1. PMP	*kaŋkaŋ	cramps, stiffening of the limbs
2. Toba Batak	baŋkaŋ dagiŋ	stiff, rigid (corpse)
3. Malay	cəkaŋ	taut, astretch
4. Malay	jəŋkaŋ	stark or stiff in death
5. Kambera	kalikaŋu	numb, stiff (as limbs)
6. Balinese	kəkaŋ	stiff
7. Malagasy	róhana	syphilitic rheumatism
8. DPB	tərkaŋ	stiff, inflexible

*-kap		to grope, feel in the dark, etc.
1. PAN	*kapkap	to feel in the dark; grope
2. Ilokano	aríkap	to feel with hand or finger; touch; fondle
3. Amis	hakap	to touch lightly, intending to find out the identity of an object
4. Hiligaynon	híkap	to touch with the hand, feel
5. Aklanon	hulíkap	to pet, feel around (the body)
6. Bontok	kapókap	to feel one's way, as a blind person
7. BM	kekap	to grope with hands and feet (in the dark, as in seeking the way)
8. BM	kokap	to grope, feel (as in the dark)
9. BM	lekap ~ likap	to grope with hands or feet, as a blind person, or someone in the dark
10. Ngadha	ŋoka	to feel in the water for fish
11. Puyuma	paLukap	to feel, finger s.t.
12. Kavalan	pRapkap	to grope
13. Buruese	raka-h	to grope for, feel for
14. Tae'	salaŋkaʔ	to grope for s.t., as a blind person
15. Maranao	sikap	to search, find, examine, grope for
16. Kavalan	sukap	to grope for fish, shrimp or crabs under a stone in the water
17. Kelabit	tekap	to search

*-kaq₋₁		to open forcibly (dbl. *-ka, *-kaq₋₂)
1. PMP	*bikaq	to open forcibly, force apart (as eyelids with fingers)
2. PWMP	*bukaq	to open, untie
3. PAN	*hukaq	to loosen, open
4. PMP	*qu(ŋ)kaq	to loosen, pry open
5. PMP	*rakaq	to open
6. PMP	*re(ŋ)kaq	to force open
7. Aklanon	baeáŋkaʔ	to take apart, dismantle

(continued)

8. Bikol	bikaʔkáʔ	having the legs spread apart
9. Rotinese	huka	to open
10. Uma	-huŋkaʔ	to open
11. Singhi	kuka	to undo, unfasten
12. Maranao	ləkaʔ	to open
13. Tboli	lemkaʔ	to take off, open forcibly
14. Rembong	léŋkaʔ	to open (a door); opened
15. Tiruray	luŋkaʔ	to loosen slightly, as a bandage
16. Ilokano	pekká	to break, shatter; open forcibly (as someone's fist)
17. Rotinese	sika	to open s.t. wide, hold s.t. open
18. Sasak	toŋkaʔ	to open (door)

*-kaq$_2$		to split, crack open (dbl. *-ka*-kaq$_{-1}$)
1. PMP	*bakaq	to spread apart, split
2. PMP	*bekaq	to split, crack open
3. PMP	*bikaq	to split
4. PWMP	*buŋkaq	to split apart; piece split off
5. PMP	*kakaq	to split
6. PMP	*laŋkaq	stride
7. PMP	*le(ŋ)kaq	to split open, as dry ground
8. PWMP	*lu(ŋ)ka	to split, cleave
9. PWMP	*pekaq	to split
10. PMP	*pu(ŋ)kaq	to break off
11. PMP	*wakaq	to split
12. Bikol	bikaʔkáʔ	having the legs spread apart
13. Javanese	caŋkah	forked branch
14. Malay	cəkah	rift, split; to split open by pressure
15. DPB	daŋkah	fork, branch
16. Buruese	geka-h	to split, as by heat
17. Bare'e	jeka	to split open by itself, as a ripe fruit
18. Iban	kərəkah	crack, fissure
19. Javanese	pləkah	to crack, split
20. Malay	rəkah	to split, show a crack
21. Wolio	seka	to split, cleave

Note: Also Dairi-Pakpak Batak *bəka* 'split in two', Karo Batak, Dairi-Pakpak Batak *taka* 'split, cleave' (both < *-ka). Possibly identical to *-kaq$_{-1}$, although the two sets show several distinguishing characteristics.

*-kas.₁		to begin
1. PWMP	*belekas	to begin
2. Balinese	aŋkas	to be always just about to do s.t.; be always prevented from doing what one intends
3. Maranao	bəkas	introduction, preface; begin activity
4. Maranao	gəkas	to begin; opening remark, preface
5. DPB	jəŋkas	very early morning; also said of a person who anticipates what s.o. else is about to say
6. WBM	lakas	to try out a new article for the first time
7. Balinese	ləkas	to begin, be going to do; beginning
8. Bare'e	poŋka	beginning, introduction
9. Madurese	poŋkas	beginning

*-kas.₂		to loosen, undo, untie
1. PMP	*bakas	to loosen, undo, untie
2. PPH	*balukas	to loosen, release the bonds
3. PAN	*bekas.₁	to spring a trap
4. PMP	*beŋkas	to untie, undo
5. PMP	*biŋkas	to undo, untie, spring a trap
6. PAN	*bukas	to release, undo
7. PWMP	*hekas	to release, loosen, untie
8. PWMP	*hukas	to loosen, untie, undress; to separate
9. PMP	*kaskas	to loosen, untie
10. PMP	*lekas.₁	to open, undress, remove, release
11. PMP	*lu(ŋ)kas	to remove, loosen, untie
12. PWMP	*pukas.₂	to release, loosen, untie
13. PMP	*taŋkas	to loosen, untie
14. PMP	*ti(ŋ)kas	to lift off a cover, be lifted off
15. PWMP	*tukas	to open
16. PMP	*wakas	to loosen, undo, uncover
17. Maranao	bokakas	to loosen, uncoil
18. Fijian	ceka	to untie a bundle
19. Hawu	deka	to peel off husk of areca nut
20. Wolio	doŋka	to get loose, slip, away, coming off (e.g. a button)
21. Maranao	gikas	to disperse, as in crowd, give way
22. Tagalog	himúkas	removal of small fish stuck in net
23. Tagalog	ʔigkás	being set off or freed, as a spring suddenly let loose or a trigger discharged
24. Iban	iŋkas	to remove, undo, unpick (stitches)
25. Sangir	kokaseʔ	to remove the bandages and wrappings of s.t.

(continued)

26. Manggarai	lahas	to untie, unravel
27. Kayan	lemkah	to release a trigger
28. Manggarai	léŋkas	to open, break open
29. Cebuano	líbkas	for a trap to spring, for s.t. held back to be released suddenly
30. Maranao	mokas	trigger, spring
31. Palawano	pakas	sprung (spring of a trap), released
32. WBM	pəkas	to slip s.t. off, as a loop off a peg, a button out of a button-hole; force off someone's grip on something
33. Wolio	piŋka	to become dislodged, detached
34. DPB	pulkas	to open; broken, of a dam
35. Bontok	ʔabkas	to spring, release or set off, of a spring release bird or animal trap
36. Ilokano	rekkás	to loosen, undo, untie, unfasten
37. Malay	roŋkas	taking to pieces; dismantling
38. Wolio	soka	to unfasten, undo, untie, break off
39. DPB	talkas	free, of a person who was previously detained; loosened, of clothes that are removed
40. Pazeh	tilikat	trap, snare in general
41. Maranao	toŋkas	to separate, put apart
42. Bikol	ʔukás	to remove or take off the clothes
43. Ngadha	xaka	to loosen, remove, detach, peel off

*-kas.$_3$		quick, agile, strong, energetic
1. PAN	*alikas	quick, fast
2. PMP	*bakas	swift, strong, energetic
3. PAN	*bekas.$_2$	swift, fast
4. PWMP	*bikas	strong, vigorous, energetic
5. PMP	*cekas	quick, swift
6. PPH	*kaskás	swift, fast, speedy
7. PMP	*lakas	quick, energetic, strong
8. PWMP	*lekas.$_2$	quick, quickly
9. PMP	*rikas	quick, fast
10. PMP	*ta(ŋ)kas	agile, quick, swift, energetic
11. PMP	*ti(ŋ)kas	swift, energetic
12. Malay	akas	dexterous; neat or quick in action
13. Sundanese	binəkas	clever, sharp, ingenious
14. Iban	caŋkas	fit, proper, strong, young
15. Lauje	iŋkas	quick, fast
16. Kankanaey	laskás	alert; agile; active; nimble; brisk; lively; quick

(continued)

17. Sangir	liŋkaseʔ	energetic; a quick and strong man
18. Ngaju Dayak	mantekas	quick, speedy, vehement
19. Dumagat (Casiguran)	ʔohikás	quick, lively (of a child)
20. Madurese	rekas	quick
21. Iban	bə-tələŋkas	moving quickly (travelling light)

*-kat		to rise, climb
1. PWMP	*aŋkat	to lift, raise, pick up
2. PPH	*kalatkat	to climb up, as a vine
3. PWMP	*katkat	to lift, raise up
4. PAN	*sakat	to rise, climb up
5. Sundanese	cəŋkat	to raise the head; rise up from a lying position
6. Dumagat (Casiguran)	ʔəgkát	to stand up, to lift up, to get up from a sitting position
7. Banggai	iŋkat	to ascend; be raised
8. Sundanese	jeŋkat	jumping up, as from a sitting position
9. Ayta Abellen	lakat	to climb, go up
10. Bikol	pákat	adept or nimble at climbing
11. Maranao	səkat	to elevate, step up stairs or rung of ladder
12. Dumagat (Casiguran)	sikát	to rise, of the sun
13. BM	tompikat	rope on the feet to help climb trees
14. Kenyah	ukat	to climb

*-kaw$_1$		a curve; to curve, meander; circuitous
1. PAN	*likaw	curve, bend, winding
2. PMP	*piŋkaw	bent, crooked
3. Pangasinan	basíkaw	curve; archway
4. Aklanon	pakaw	twisted, deformed (of the arm)
5. Isneg	pulikawkaw	undulatory, curved lines

*-kaw$_2$		high, tall (dbl. *-baw$_1$)
1. PMP	*laŋkaw	high, tall
2. Tiruray	akaw akaw	a pair of stilts; to walk on stilts
3. Kayan (Uma Juman)	boŋko	upper part, upper surface
4. Ilokano	daŋkáw	long-legged

(continued)

5. Ilokano	kawákaw	very tall, much taller than the generality of his companions
6. Miri	kew	tall
7. Tiruray	saŋkaw	to jump up to grasp
8. Binukid	taŋkaw	tall in height, high (object)
9. Cebuano	tukáw	to tower over

*-keb.₁		lid, cover; to cover (dbl. *-kub.₁, *-kep.₁, *-kum, *-kup)
1. PWMP	*aŋkeb	covering, lid
2. PMP	*e(ŋ)keb	a cover, to cover
3. PMP	*kebkeb	to enclose, shut in
4. PWMP	*lekeb	to cover; shut in
5. PWMP	*lu(ŋ)keb	to cover, shelter
6. PMP	*Ruŋkeb	to cover over, cover up
7. PPH	*salakeb	cover trap for fish or crustaceans
8. PWMP	*sa(ŋ)keb	cover, lid
9. PWMP	*se(ŋ)keb	cover
10. PWMP	*si(ŋ)keb	cover
11. PWMP	*ta(ŋ)keb	a cover, overlapping part
12. PWMP	*tekeb	to cover
13. PWMP	*tikeb	to close up
14. Javanese	dəkəb	to catch by covering with the hands (as a bird)
15. Maranao	gikəb	walled tightly
16. Dusun (Kadazan)	hikob	to fold
17. Tiruray	laŋkəb	cage trap
18. Sundanese	riŋkəb	closed
19. Paiwan	tsukev	a large cover (as for cookpan)

*-keb.₂		to lie face down be prone (dbl. *-kem-₂)
1. PAN	*kebkeb	to lie prone, face down
2. PWMP	*la(ŋ)keb	to lie face down
3. PWMP	*le(ŋ)keb	to lie prone, face down
4. PPH	*sakeb	to lie face down, prone
5. PWMP	*taŋkeb	to fall face down
6. PWMP	*tiŋkeb	to lie prone, face down
7. Maranao	galəkəb	face side down, tails in coin
8. Agta (Eastern)	hákab	to lie face down
9. Balinese	kakəb	lie forward, fall forward
10. Ngaju Dayak	mahiŋkep	to lie on the stomach
11. Javanese	ruŋkəb	to fall face down

(continued)

12. Hanunóo	súkub	facing down, pronation
13. Maranao	taləkəb	to lie face down
14. Gorontalo	toŋgobu	overturned
15. Javanese	uŋkəb-uŋkəb	to lie face down

Note: Also PPH *lukub 'to lie facing down', Tagalog *taŋkáb* 'to fall, hurting one's lip or chin'.

*-keC	adhesive, sticky	
1. PMP	*buket	thick (of fluids), viscous
2. PMP	*ca(ŋ)ket	sticky; to stick to
3. PMP	*da(ŋ)ket	to stick, adhere
4. PPH	*dayket	sticky, glutinous, adhesive
5. PAN	*dekeC	paste, adhesive (probably made of overcooked rice)
6. PAN	*dikeC	sticky, adhesive
7. PWMP	*e(ŋ)ket	to stick, adhere
8. PMP	*la(ŋ)ket	to stick, adhere; sticky, viscous
9. PWMP	*leket	to stick, adhere
10. PMP	*liket	sticky, adhesive
11. PMP	*maket	to stick, adhere to; sticky, adhesive
12. PMP	*niket	sticky, adhesive
13. PMP	*ña(ŋ)ket	sticky, adhesive
14. PMP	*peket	sticky, adhesive
15. PWMP	*piket	sticky, adhesive
16. PWMP	*piReket	sticky, adhesive
17. PWMP	*puket	to stick, cling to
18. PMP	*raket	sticky, adhesive
19. PWMP	*reket	to stick, adhere to; sticky, adhesive
20. PMP	*riket	sticky
21. PWMP	*seket	to stick, adhere; glue
22. PMP	*si(ŋ)ket	to stick, adhere to
23. PWMP	*ta(ŋ)ket	to adhere, stick to
24. PWMP	*zi(ŋ)ket	to adhere, stick to
25. DPB	alkət	viscous
26. Pangasinan	ʔansakkə́t	sticky rice
27. Kankanaey	busíkət	clay; the thick, gluey black earth of rice fields
28. Ayta Abellen	dalikət	honey; sticky rice
29. DPB	darkət	to be parasitic on
30. Ifugaw	dayakót	general term for sticky rice
31. Lun Dayeh	duket	to stick on or stick with
32. Cebuano	hágkut	sticky
33. Ngaju Dayak	hiket	close together

(continued)

34.	Tiruray	kakət	(of rust) to adhere to a rusted object
35.	Tagalog	lagkít	stickiness
36.	Javanese	lukət	close, intimate
37.	Mansaka	masaŋkət	adhesive, sticky
38.	Palauan	merekeréked	sticky
39.	Tboli	meket	thick, sticky, as the consistency of cooked cereal or corn soup
40.	Teluti	miket	sticky
41.	Palauan	omeréked	paste, glue
42.	Kankanaey	paŋkə́t	sticking, adhering, cleaving
43.	Kenyah (Long Wat)	pəlikət	sticky, adhesive
44.	Ifugaw	punikót	stickiness
45.	Ilokano	puríket	grass with thorny appendages that stick to clothing: *Bidens pilosa*
46.	Sika	ʔaket	to stick, remain somewhere
47.	Bikol	ragkót	nonslip (the opposite of slippery)
48.	Palauan	rekeréked	sucker fish
49.	Javanese	rukət	to stay close to
50.	WBM	rumikət	sticky (as honey, mud)
51.	Tiruray	sarəkət	glue, to glue
52.	Amis	sikət	attached to, joined to
53.	Itbayaten	ma-solichət	sticky (mud, rice, food, etc.)
54.	Lamaholot	tərəkət	to paste on, stick to, glue
55.	Rejang	tikit	tree with sap used for bird lime
56.	Pangasinan	yakə́t	sap

*-ked		prop, support; staff (dbl. *-kad, *-kud)
1. PAN	*suked	a prop, support; to prop up or support
2. PMP	*teked	to prop, support
3. PAN	*tuked	to prop, support
4. Madurese	duŋket	staff, walking stick
5. Tiruray	fakər	brace, support
6. Ilokano	sammakéd	to prop, support
7. Ilokano	sarikedkéd	foundation, basis, prop, support
8. Isneg	síkād	to stand up; put one's foot against some support (for exertion)
9. WBM	sugkəd	walking stick; to use a walking stick for support
10. Ilokano	talkéd	to lean, rest on for support (Carro 1956)
11. PNV	*tiko	walking stick, canoe pole

*-kek		to shriek, creak, cluck, chuckle (dbl. *-kak, *-kik, *-kuk₂)
1. PMP	*e(ŋ)kek	subdued laughter; sobbing
2. PCEMP	*gekek	to snort, gasp
3. PMP	*kek	to squawk
4. PMP	*ke(k)kek	to cluck
5. PMP	*tekek	dull throaty sound
6. PMP	*tikek	cry of the gecko
7. Iban	cəkak	to seize in the hands, throttle
8. Sundanese	cəkakək	the sound of laughter
9. Tetun	fokok	nasal, speaking through the nose
10. Ifugaw	pakók	cackling sound
11. Malay	səŋkak	nausea causing hiccoughs
12. Dusun (Kadazan)	sikok	to sob

*-kel.₁		to bend, curl
1. PAN	*Cikel	to bend
2. PMP	*eŋkel	bent, stooped
3. PWMP	*ikel	curly, wavy (as the hair)
4. PMP	*pikel	bend, curve
5. PAN	*sekel	to bend, bow
6. Banggai	bokol	wave, billow
7. Balinese	diŋkəl	crooked, bent over
8. Keley-i	kəlkəl	a type of necklace
9. Fijian	roko	to be in a bowing or stooping position, a sign of reverence
10. Bare'e	seŋko	bent, bowed
11. Balinese	takəl	doubled, folded
12. Manggarai	tekel	to coil up, as a snake
13. Bare'e	teŋko	bent, twisted
14. Ifugaw	túkol	huddling position of a chicken or bird
15. Javanese	ukəl	in the shape of a coil

Note: Also Balinese *iŋgəl* 'curl, be curly; twisted; coil'.

*-kel.₂		harsh coughing or choking
1. PMP	*cekel	to choke, strangle
2. PWMP	*kelkel	convulsive, of coughing or laughing
3. Tausug	buŋkul	to get a lump of food caught in the esophagus
4. Puyuma	ḍekeḷ	to choke on s.t. stuck in one's throat
5. Tausug	kagulkul	the sound of harsh coughing
6. Toba Batak	oŋkol	cough
7. Maranao	pəkəl	to choke, strangle

(continued)

8. Tausug	pikul	to squeeze the neck with the purpose of choking, choke s.o.
9. Ibaloy	sikel	to have s.t. stuck in one's throat, "go down the wrong way"
10. Old Javanese	waŋkəl	to get stuck (in the throat, etc.)

Note: Also Casiguran Dumagat *sakál* 'to strangle, choke s.o.', Kalamian Tagbanwa *kulkul*, Agutaynen *keykey* 'cough; to cough'.

*-kem.$_1$		to clench, cover; grasp (dbl. *-gem)
1. PWMP	*buŋkem	to close the mouth
2. PWMP	*ca(ŋ)kem	grasp, hold
3. PWMP	*iŋkem	to close, shut
4. PPH	*kem	clenching, gripping, grasping
5. PWMP	*kemkem	to hold in the fist
6. PMP	*qeŋkem	to enclose, grasp
7. PMP	*Re(ŋ)kem	to grasp in the fist
8. PPH	*sakém	to grasp, grasp for; grasping, greedy
9. PWMP	*takem	tightly closed
10. PWMP	*ti(ŋ)kem	closed, shut (as the mouth)
11. Chamorro	akihom	to clench, grip
12. Tontemboan	aŋkəm	to take in the hand
13. Javanese	bəkəm	to clutch in the hand
14. Javanese	blaŋkəm	(of the mouth) not having been used (for talking, eating) for hours on end
15. Balinese	dəkəm	to hold firmly, hold fast to
16. Dusun (Kadazan)	gakom	to catch, capture, seize, arrest
17. Ifugaw	gokóm	fist, clenched hand
18. Hiligaynon	hugakúm	to clasp in the fist
19. Dusun (Kadazan)	kakom	grasp
20. Ayta Abellen	kayəmkəm	to close the fist
21. Malay	ləkam	to grasp in the hollow between the forefinger and thumb
22. Pazeh	mohakem	to embrace
23. Kenyah	ñekem	to grasp, enclose with the fist
24. Maranao	sikəm	snatching with the hand
25. Javanese	təkəm	a handful

*-kem.$_2$		to lie face down, be prone (dbl. *-keb-$_2$)
1. Manggarai	eŋkem	to overturn
2. Karo Batak	laŋkəm	lie on the stomach

(continued)

3. Manggarai	mekem	prone, face-downwards
4. Old Javanese	suŋkəm	to bend forward, lie face-down
5. Kayan	tukem	facing downwards

Note: Karo Batak *laŋkəm* might reflect *-keb-$_2$ 'to lie face down, be prone'.

*-keŋ-$_1$		cramps, stiffening of limbs
1. PMP	*e(ŋ)keŋ	stiff, as a corpse
2. PMP	*keŋkeŋ	cramps, stiffening of limbs
3. PWMP	*pekeŋkeŋ	stiff, of the body
4. Javanese	bəkəŋkəŋ	unyielding
	dandan bəkəŋkəŋ	to dress in stiffly ironed clothes
5. Javanese	bəŋkəŋ	(of muscles) stiff and sore
6. Javanese	cəkəŋkəŋ	stiffening, becoming rigid, as during a paralytic stroke
7. Malay	chəkaŋ	taut, stretch
8. Malay	jəŋkaŋ	stark or stiff in death
9. Javanese	rəŋkəŋ	to get up with difficulty, because of a debilitating illness, or sore muscles
10. Malagasy	vónkina	contracted, drawn up (as the limbs)

Note: Also Tonbomuwo *o-kikaŋ* 'stiff'.

*-keŋ-$_2$		hollow, resounding sound (dbl. *-kaŋ-$_1$, *-kiŋ, *-kuŋ-$_2$)
1. PMP	*eŋkeŋ	to moan, hum, howl
2. PMP	*keŋ	hollow, resounding sound
3. PAN	*keŋkeŋ	hollow, resounding sound
4. Malay	bəlaŋkaŋ	to ring (of metal falling on stone)
5. Ifugaw	bíkoŋ	lover's harp; flat, thin metal plaque, the inner, loose part of which can vibrate and produce harplike sounds
6. DPB	dəkaŋ	sound made by the rusa deer (*Cervus equinus*)

*-keŋ-$_3$		to shrink, shrivel
1. PMP	*keŋkeŋ	to shrink
2. PMP	*Reŋkeŋ	to shrink, contract
3. Malagasy	ónkina	to shrink, be shrivelled
4. Fijian	saqoqo	shrivelled, as a leaf; shrunk, as a cloth
5. Malagasy	vónkina	to contract, shrivel, shrink

*-kep₁		to cover; fold over (dbl. *-keb₁, *-kub₁, *-kum, *-kup)
1. PWMP	*cakep	to close, shut
2. PMP	*ekep	to brood, sit on eggs
3. PWMP	*lekep	to cover, shut in
4. PPH	*liŋkep	to close up, shut (as a door)
5. PPH	*liqekep	to shut, close, cover
6. PWMP	*ri(ŋ)kep	a cover; to cover
7. PWMP	*sa(ŋ)kep	to close, shut
8. PMP	*sekep	to keep secret, cover up
9. PAN	*takep	to fold over, double
10. PAN	*tekep	a cover, to cover with a flat surface
11. Balinese	aŋkəp	double, to fold double
12. Maranao	kakəp	to surround, encircle, close
13. Maranao	laŋkəp	a patch, thin metal strip
14. Sasak	luŋkəp	to cover, cover up
15. Iban	rəŋkap	to cover, put over
16. Puyuma	salikəp	to close a door
17. Javanese	siḍakəp	with arms folded across the chest
18. Samoan	siʔo	enclosed; to surround, encircle
19. Iban	suŋkap	to put or slip under
20. Puyuma	tatəkəp	booby trap (wooden plank with bait, and bamboo cover that closes on the animal)
21. Palawano	tiŋkep	lid, cover; basket with a lid
22. Dusun (Kadazan)	tukop	to cover up for another's wrong action
23. Singhi	ŋ-ukop	to sit (of a hen)

*-kep₂		to seize, grasp, embrace
1. PWMP	*cakep	to close, shut
2. PMP	*cekep	to seize, grasp
3. PMP	*cikep	to catch with the hands
4. PMP	*dakep	to seize, arrest
5. PAN	*dekep	to embrace
6. PWMP	*hakep	to grasp, embrace
7. PMP	*kepkep	to seize, hug, embrace
8. PWMP	*Rakep	to seize, grasp, wrap around
9. PMP	*sakep	to catch, seize
10. PWMP	*sikep	hawk sp.
11. PMP	*ta(ŋ)kep	to seize, catch
12. PWMP	*tikep	to catch, seize
13. Old Javanese	dikəp	to catch with the flat of the hand
14. Motu	dogo-a	to seize, take hold of; restrain
15. Sasak	dukəp	to catch, grasp with fingers or claws
16. Tiruray	ʔəkəf	to embrace, cling
17. Ida'an Begak	gakop	to put arms around

(continued)

18. Tboli	lekef	to hug, embrace
19. Fijian	moko	to embrace, clasp in the arms
20. Iban	pakap	to put the arms round, embrace
21. Puyuma	rəkəp	falcon, hawk
22. Dumagat (Casiguran)	salikə́p	peregrine falcon
23. Balinese	sidakəp	folding the arms on the chest
24. Amis	təkəp	kind of trap
25. Sasak	tukəp	to seize, grasp, catch

*-keR		stiff, rigid, as a corpse (dbl. *-teR)
1. Binukid	bisekeg	stiff (as abaca fiber)
2. Sundanese	jəŋkər	to be very stiff, as a dead body
3. Manggarai	leŋker	stiff (of the body); rigid (of a corpse)
4. Maranao	pamikəg	stiff, stiffen; rigor mortis
5. Bikol	taliskog	a splint, cast (as for a broken arm); to place one thing within s.t. else so that it becomes rigid
6. Cebuano	tískug	stiffen so as to lose a pliant quality

Note: Also Tiruray bəkar 'stiff, sturdy', Tontemboan tə'kər 'stiff, rigid; numb'.

*-kes		to encircle, wrap firmly around (dbl. *-kis$_{-2}$, *-kus)
1. PAN	*bakes	belt; anything that encircles tightly
2. PPH	*balíkes	to encircle, wrap around
3. PMP	*beRkes	bundle (as of firewood), package
4. PPH	*hag(e)kes	to wrap around; to hug
5. PMP	*i(ŋ)kes	bound firmly
6. PWMP	*keskes	to wrap tightly around
7. PMP	*lekes	to roll, curl up
8. PMP	*likes	to wrap firmly around
9. PPH	*tag(e)kés	to wrap tightly around
10. PAN	*takes	to wrap around; embrace, hug
11. PWMP	*tekes	to bind firmly
12. PWMP	*ti(ŋ)kes	to wind around, encircle
13. Maranao	bəkə-bəkəs	to bind well and strongly; band-like
14. WBM	bərakəs	to encircle a number of objects by the arm and hand or by the thumb and fingers
15. Karo Batak	biŋkəs	a package of salt
16. Ayta Abellen	bitkəh	to strangle
17. Maranao	gakəs	to embrace, hug

(continued)

18. Hanunóo	hákus	to hug, embrace
19. Tagbanwa (Central)	lagkɨt	to tie, fasten, or bind with a rope
20. Keley-i	lakəh	to bundle things together, as cane, wood, or branches
21. Muna	lunko	to wrap a sarong closely around the body
22. Singhi	mokus	to tie
23. Tombonuwo	oŋkos	a bandage
24. Abaknon	pakkos	to bind, tie; bundle
25. Iban	perekas	a bundle (as of sticks)
26. Puyuma	rəkəs	to tie together (sheaves, bundles)
27. Balinese	riŋkəs	wrap up, tie together, roll up (in cloth)
28. Hiligaynon	wágkus	belt, sash, waist band

*-kik		shrill throaty sound (dbl. *-kak, *-kek, *-kuk.$_2$)
1. PMP	*e(ŋ)kik	shrill cry
2. PMP	*kik	to peep, squeak, giggle
3. PMP	*kikik	to squeak, as a mouse; to giggle
4. PMP	*pikik	to chirp, squeak
5. PWMP	*qi(ŋ)kik	to squeak, giggle
6. PWMP	*re(ŋ)kik	to have a squeaky voice
7. PWMP	*tekik	high-pitched sound
8. Old Javanese	aŋkik	panting, gasping
9. Javanese	bəkak-bəki:k	to shriek repeatedly
10. Malay	bələkek	a bird, the snipe
11. Karo Batak	bəŋkik	fruit bat, flying fox
12. Ilokano	bukík	asthma
13. Malay	cəkek	seizing the neck, throttling; choking with food or guzzling
14. Sundanese	cikikik	howling, of laughter
15. Javanese	lékék	to laugh in high-pitched tones
16. Iban	ləkik	to giggle
17. Old Javanese	ŋikik	shouting with jeering laughter (of demons.)
18. Ilokano	okík	the cry of the fruit bat, or flying fox
19. Malay	pəkek	a shrill cry, shriek of pain
20. Sundanese	rakek	cry of small animals (as puppies)
21. Kambera	rékiku	to cough, clear one's throat
22. Malay	riŋkek	the neigh of a horse
23. Maranao	tarkik	to laugh heartily or vigorously
24. Malay	toʔkek	gecko

*-kiŋ		a clear ringing sound (dbl. *-kaŋ₋₁, *-keŋ₋₂, *-kuŋ₋₂, *liŋ₋₁, *-tiŋ)
1. PAN	*dekiŋ	to bark, of a deer
2. PMP	*kiŋ	ringing sound
3. PAN	*kiŋkiŋ	ringing sound
4. Balinese	caŋkiŋ	a high-pitched voice
5. Malay	cəŋkiŋ	yelp (of dog); scream (of deer)
6. Maranao	gariŋkiŋ	to ring, tinkle
7. Manggarai	ŋkiŋ	ring, sound of a bell
8. Ngaju Dayak	sakiŋkiŋ	continuous barking
9. Karo Batak	səlkiŋ	to shout loudly in the ear of a deaf person
10. Iban	təkiŋ	sound of tapping, as on a plate to call a cat
11. Nggela	tiŋgi	to ring a bell

*-kiq		a high-pitched vocal sound
1. PMP	*leŋkiq	a scream; to scream
2. PPH	*piŋkiq	to strike two hard objects together to produce a spark or sound
3. PAN	*tekiq	high-pitched sound, chirp of a gecko
4. PPH	*tikiq	gecko
5. Balinese	biŋkih	to gasp, pant, sigh
6. Bahasa Indonesia	ter-kikih	to giggle
7. Maranao	səŋkiʔ	to sob, cry
8. Bikol	tagaŋkíʔ	the sound of two pieces of iron being hit together; to clink

*-kis₋₁		to scratch, grate, scrape (dbl. *-gis)
1. PPH	*kaRiskis	to scrape, scratch
2. PAN	*kiskis	to shave, scrape off
3. Ngaju Dayak	ikis	to scrape, grate
4. Ida'an Begak	kigkis	to scrape clean; broke, without money
5. Bontok	kisakis	to scratch continually, as when bitten by bedbugs
6. Rembong	rakis	to scratch with the nails
7. Iban	tukis	to scratch, score, make a shallow cut
8. Kankanaey	uákis	to line, mark, groove

*-kis.₂		to tie, bind, wrap around (dbl. *-kes, *-kus)
1. PPH	*balíkis	to tie around; belt
2. PWMP	*beRkis	to bind up; bundle
3. PWMP	*biŋkis	to bind, tie together
4. PEG	*liŋkiso	bracelet
5. Tiruray	bərikis	to tie all over so that the one tied cannot move
6. Aklanon	bígkis	sash (piece of cloth wrapped around the waist)
7. Manggarai	ikis	to tether, tie, bind
8. Balinese	lukis	lump of rice-porridge wrapped in a leaf (for a journey)

*-kit.₁		to bite
1. PMP	*a(ŋ)kit	to bite
2. PWMP	*i(ŋ)kit	to bite
3. PWMP	*kitkit	to bite
4. WBM	kiʔit	to bite off s.t. with the front teeth
5. Soboyo	koki	to bite
6. Kambera	pàŋgitu	to chew on
7. Ngaju Dayak	paŋkit	to bite
8. Malay	saŋkit	to chatter, of the teeth
9. Gayo	sikit	to bite
10. Kelabit	uʔit	upper incisors
11. Kambera	wiŋgitu	to bite

*-kit.₂		to join along the length
1. PPH	*biŋkit	joined along the length
2. PMP	*dakit	to join along the length; raft
3. PMP	*de(ŋ)kit	touching or joined along the length
4. PWMP	*diŋkit	touching or joined along the length
5. PWMP	*liŋkit	to stick, adhere; join together
6. PMP	*pi(ŋ)kit	sticking to, attached to, joined
7. PWMP	*ri(ŋ)kit	touching or joined along the length
8. PMP	*Rakit	to lay long objects side by side; raft
9. PMP	*sa(ŋ)kit	to tie, fasten together
10. PWMP	*ziŋkit	touching or joined along the length
11. Ayta Abellen	akkit	to join with
12. Malay (Jakarta)	belénkét	adhering to one another, esp. of the eyelids
13. Fijian	buki	to tie, fasten

(continued)

14. Javanese	cukit	chopsticks; bamboo sticks used (in pairs) for mixing tobacco and opium for smoking
15. Ilokano	dalákit	raft
16. Agutaynen	dilkit	for parts of the body to rub together at a joint (armpits, thighs, neck, etc.)
17. Cebuano	íkit	close together, usually people
18. Cebuano	íŋkit	fingers which are joined congenitally together
19. Javanese	kékét	to stick tight, hold together firmly
20. Cebuano	laŋgíkit	to tie or attach things together or side by side
21. Tausug	laŋkit	adjoining, adjacent, neighboring
22. Malay	ləkit	adhering lightly
23. Abaknon	nikit	attached, joined
24. Bare'e	raŋki	what one puts next to, or takes together with s.t. else
25. Maranao	rəkit	to connect, join
26. Tiruray	sarikit	a double-thumb deformity
27. Cebuano	síŋkit	to tie two coconuts together by their husks partly stripped off

*-kit.$_3$ to remove, detach (dbl. *-nit)

1. PPH	*úkit	to remove s.t. from an attached position
2. Keley-i	akit	to remove the core of a log or the trunk of a tree
3. Keley-i	duʔkit	to remove a small wedged item using a screwdriver or other sharp tool
4. Bikol	puʔkít	to detach or disconnect
5. Ibaloy	sokit	to extract, remove s.t. from a hole or a container with an instrument

*-kit.$_4$ small

1. PAN	*dikit	little, few, small in amount
2. PMP	*kikit	small, trifling
3. Itneg	battikit	small
4. Ibaloy	esokit	too small (to accommodate s.t.)
5. Aklanon	ikít	close, narrow
6. Bare'e	toki	small

*-kub.₁		a cover; to cover (dbl. *-keb.₁, *-kep.₁, *-kum, *-kup)
1.	PMP	*aŋkub	cover
2.	PWMP	*cu(ŋ)kub	to cover with a hollow container
3.	PMP	*kubkub.₂	cover, coating
4.	PWMP	*Raŋkub	to cover
5.	PPH	*sakúb	lid, cover
6.	PPH	*salúkub	to cover s.t. to protect it or oneself
7.	PWMP	*sekub	a cover; to cover
8.	PWMP	*su(ŋ)kub	to cover with a hollow container
9.	PMP	*ta(ŋ)kub	to close or cover up
10.	PMP	*tekub	to cover, shut
11.	PWMP	*tu(ŋ)kub	cover
12.	PWMP	*ukub	cover
13.	Aklanon	eáŋkob	to join together, cover a gap
14.	WBM	kəlubkub	to add a cover to keep warm
15.	Javanese	krukub	shade, fabric covering
16.	Sangir	lakubə?	to overhang, and so overshadow
17.	Sangir	liŋkubə?	to cover one thing with another
18.	Maranao	lokob	lid, cover
19.	Ifugaw	lokúb	patch used to cover a hole
20.	Kavalan	nukub	to cover
21.	Lun Dayeh	rukub	a protection or shelter
22.	Ilokano	talákub	reglets used to cover joints between boards

*-kub.₂		to surround, encircle
1.	PPH	*kubkub	to surround, lay siege to
2.	PPH	*lakub	enclosure; to enclose
3.	PWMP	*li(ŋ)kub	to surround, encircle
4.	PPH	*lukúb	to encircle, surround
5.	Ibaloy	alikobkob	to be surrounded by mountains, as a village in a valley
6.	Ibaloy	dekob	to corner s.t., to surround s.t.—of the action of many people
7.	Ilokano	íkub	enclosure for freshly reaped rice
8.	Manobo (Sarangani)	kelokob	pen for animals
9.	Kankanaey	salikubkúb	to surround, encircle, inclose
10.	Kankanaey	sókub	to sit, etc. in a ring
11.	Kankanaey	tikúb	to hedge in, fence in, encircle
12.	Dusun (Kadazan)	tuŋkakub	trap for birds

*-kuC		hunched over, bent (dbl. *-kug, *-kuk₋₁, *-kul₋₁, *-kuŋ₋₁, *-kuq, *-kux, *-luk/luquk, *-tuk₋₂)
1. PPH	*balikutkút	to bend, stoop, curl
2. PAN	*bekuC	to bend over (of a person); hunchbacked
3. PMP	*bikuC	hunched over
4. PMP	*bukut	hunchbacked
5. PAN	*kuCkuC	to roll up (paper), curl up tightly (body)
6. PMP	*lekut	bent, as a limb
7. PMP	*likut	curled up
8. PMP	*lukut	to roll or crumple up
9. PMP	*pikut	to bend over (as a person)
10. PAN	*qukut	hunched over, bowed (as with age)
11. PMP	*rekut	curled or crumpled up
12. POC	*ruku	to stoop, bow
13. PMP	*sikut	bent, of the arm
14. Lun Dayeh	belakut	bent; to bend
15. Ifugaw	bikutkút	hump in the back of old people
16. Mansaka	boyokot	to double up the body
17. Manggarai	dekut	to bow the head
18. Amis	fickot	to cramp, double up with pain; have a spasm, convulse
19. Teluti	heŋkut	to roll up loosely
20. Tagalog	hukót	bent-backed
21. Ngaju Dayak	jukot	bent, hunched over (people, animals)
22. Kenyah	lakut	bent over, as a person
23. Manggarai	pekut	curved, of the penis
24. Kenyah	pelakut	to bend over, as ripe rice
25. Puyuma	sekuṭ	to bend (knee or elbow)
26. Tagalog	súkot	to cower, go away with head bowed in shame or humiliation

*-kud		walking stick; prop; support (dbl. *-kad, *-ked)
1. PR	*okoDo	stick, pole
2. PAN	*sukud	walking stick, cane, staff
3. PAN	*tukud	cane; prop, support
4. Keley-i	alukud	to walk beside a person to help or guide him (as someone who is drunk)
5. Lun Dayeh	rukud	walking stick
6. Puyuma	sarkuḍ	walk using a stick; walking stick
7. Ilokano	sarukód	walking stick, staff, cane
8. Kenyah (Long Wat)	sekut	walking stick
9. Itbayaten	sichod	walking stick, cane
10. Isneg	sìdukúd	cane, walking stick
11. Bontok	sol-solkod	stick or reed used as a walking stick, or to support a heavy shoulder load
12. Bontok	tolkod	house post

*-kug		curved, bend (dbl. *-kuC, *-kuk.₁, *-kul.₁, *-kuŋ.₁, *-kuq, *-kux, *-luk/luquk, *-tuk.₂)
1. PMP	*bekug	curved, bent
2. PWMP	*bikug	curved, bent
3. PWMP	*buŋkug	bent over, hunchbacked
4. PWMP	*li(ŋ)kug	curl, curve
5. PWMP	*tikug	to bend; bent, curved
6. Bare'e	aŋku	bent forward, as s.o. carrying a heavy load
7. Kankanaey	bakúg	hunchbacked
8. Cebuano	baliskúg	to become curled up, rolled
9. Kankanaey	batikúg	curved; bent; bowed; inclined
10. WBM	burigkug	a crooked cat's tail
11. Arta	dukug	to bow the head
12. Cebuano	kúgkug	to curl up stiff and hard
13. Ilokano	pilkóg	doubled, bent, curved, crooked, turned (nails, edges of knives, etc.)
14. Cebuano	síŋkug	lame, cripple
15. Maranao	təkog	to curve, bend
16. Kankanaey	yákug	hunchbacked

*-kuk.₁		bent, crooked (dbl. *-kuC, *-kug, *-kul.₁, *-kuŋ.₁, *-kuq, *-kux, *-luk/luquk, *-tuk.₂)
1. PMP	*bekuk	curved, bent
2. PMP	*biŋkuk	curved, bent
3. PMP	*buŋkuk	bent, crooked
4. PMP	*ci(ŋ)kuk	bent, curved
5. PMP	*dekuk	to bow, bend downward
6. PMP	*duŋkuk	hunched over
7. PMP	*leŋkuk	bend, curve
8. PMP	*pi(ŋ)kuk	a bend, curve (as in a road)
9. PMP	*si(ŋ)kuk	have a stiff or bent arm
10. PMP	*su(ŋ)kuk	to hang one's head
11. PMP	*tekuk	to bend or pull down, as a branch
12. PWMP	*tu(ŋ)kuk	to lower the head; bow or bend down
13. PMP	*zuŋkuk	to bend over, stoop
14. Malay	bəleŋkok	crooked; bent
15. Old Javanese	huŋkuk	hunchbacked; crooked
16. Iban	iŋkok	crooked
17. Kayan	jelukuk	to walk with bent-over posture
18. Toba Batak	laŋkuk	crooked, bent (< *-kug?)
19. Iban	ləlukok	(of old people) bent, stooped
20. Bisaya (Limbang)	liŋkuk	bent

(continued)

21. Wolio	luku	to bow, submit, give in, be meek
22. Manggarai	nikuk	corner, bend of a road; crooked
23. Manggarai	ŋgukuk	bent, bowed, humpbacked
24. Keley-i	pakuk	crooked, usually concerning wood, but may be used idiomatically regarding speech
25. Malay	pəkok	twisted or malformed (of a limb)
26. Rotinese	puku	hunched, bent, of the back
27. Amis	ʔotkok	to huddle, as when cold
28. Javanese	réŋkok	zigzagging, twisting and turning
29. Iban	rukok	bent and shambling (as an elderly person)
30. Buruese	seku-k	to bend into an angle (arm or knee)
31. Manggarai	tiŋkuk	concavity

Note: Forms in Iban, Malay, Wolio, Manggarai and Rotinese could reflect *-kug.

*-kuk.$_2$		sound of a sob, cackle, etc. (dbl. *-kak, *-kek, *-kik)
1. PWMP	*cekuk	to choke, gag
2. PMP	*dekuk	sound of a sob, grunt, etc.
3. PMP	*e(ŋ)kuk	to make a croaking sound
4. PMP	*kuk	sound of a sob, croak, etc.
5. PMP	*kukkuk	to cluck, of chickens
6. PMP	*sekuk	to sob, choke
7. PPH	*tekuk	to sound off, cry out
8. PAN	*tuRukuk	chicken
9. PMP	*ukuk	a cough; to cough; sound of coughing
10. Sika	bekuk	rustling; squeaking of mice
11. Malay	bələkok	a bird: pond heron or gallinule
12. Iban	əmpəkok	to crow, of a cock
13. Manggarai	heŋkuk	incessantly coughing
14. Javanese	klukuk	of an empty stomach, to rumble
15. Yakan	lekok-lekok	to grunt, of a pig
16. Manggarai	raŋkok	the sound made by a hen; sound made by newly hatched chicks
17. Iban	rikok	to creak, squeal
18. Iban	sikok	to sob, choking sobs
19. Kambera	takuku	crowing of a cock
20. Kavalan	teraquq	chicken
21. WBM	tuŋkuk	to tap on something lightly

*-kul₁		to curl, bend (dbl. *-kel₁, *-kuC, *-kug, *-kuk₁, *-kuŋ₁, *-kuq, *-kux*, *-luk/luquk, -tuk₂)
1. PMP	*bekul	curved, bent
2. PMP	*buŋkul	a swelling, lump or bump under the skin
3. PWMP	*deŋkul	to bend, fold
4. PWMP	*diŋkul	to bend a limb
5. PWMP	*duŋkul	to bend a limb
6. PWMP	*iŋkul	bent into a curve
7. PWMP	*kulkul	to bend the limbs, curled up the body
8. PMP	*pikul	to bend, curve
9. PMP	*siŋkul	bent, of the arm
10. PWMP	*tikul	curved, warped, as a board
11. PWMP	*tukul	to huddle, bow the head
12. WBM	bakul	to bend
13. Lun Dayeh	ekul	a roll of rope, wire, or rattan
14. Iban	əmpukul	to lie with legs drawn up
15. Amis	fitokol	dried and bent over, though still alive, as a misshapen tree
16. Sundanese	jəŋkul	the bending of the knee or other joint
17. Tolai	kăkul	curved
18. Tetun	naksekul	prominent crooked teeth
19. Old Javanese	pəkul	to embrace, hug
20. Javanese	réŋkol	full of twistings and turnings
21. Iban	suŋkul	to bend
22. Puyuma (Tamalakaw)	Tekur	to bend
23. Pazeh	tibukul	crooked, curved (*/l/ > /l/ regular?)
24. Rembong	wikul	(of women) to sit with legs folded to the side

*-kul₂		snail
1. Keley-i	basikul ~ batsikul	snail
2. Bintulu	bəkul	snail
3. Wolio	biku	snail
4. Pangasinan	bisokól	snail

*-kum		to close by folding (dbl. *-keb₁, *-kub₁, *-kep₁, *-kup)
1. PWMP	*kumkum	to fold the arms or wings
2. Tagalog	ʔikóm	closed (said of flowers and buds)
3. Iban	siŋkum	(of mouth, seed pods, etc.) closed, shut, tight
4. Tboli	stikom	to close, fold shut, as a flower; (of a wound) to heal

(continued)

5. Manggarai	teŋkum	to embrace
6. Tboli	tikom	sensitive plant: *Mimosa pudica*

*-kuŋ$_{-1}$ to bend, curve (dbl. *-kuC, *-kug, *-kuk.$_1$, *-kul.$_1$, *-kuq, *-kux, *-luk/luquk, *-luŋ.$_1$, *-tuk.$_2$)

1. PMP	*beŋkuŋ	bent; curve, arch
2. PMP	*biŋkuŋ	implement with curved blade, adze; hammerhead shark
3. PMP	*bilikuŋ	bend, curve
4. PWMP	*buqekuŋ	bend, curve
5. PMP	*ceŋkuŋ	deeply curved
6. PMP	*dekuŋ	to bend over
7. PAN	*dukuŋ	to bend, curve
8. PWMP	*e(ŋ)kuŋ	curved, crooked
9. PMP	*kuŋkuŋ	curve, curved; hollow
10. PMP	*leŋkuŋ	concave, curved inward
11. PWMP	*li(ŋ)kuŋ	concave, curving inward
12. PPH	*lukuŋ	curve
13. PWMP	*piŋkuŋ	bent, curved
14. PWMP	*sa(ŋ)kuŋ	curved inward, sunken
15. PWMP	*siŋkuquŋ	curved, having a curved edge
16. PWMP	*su(ŋ)kuŋ	stooped, bent (as the back)
17. PMP	*tekuŋ	to bow the head
18. PMP	*tikuŋ	curved, crooked, bent
19. PMP	*zeŋkuŋ	curved
20. Amis	akuŋ	crooked
21. Binukid	balikuŋ-an	a curve (in a path or road)
22. Tae'	balekoŋ	grown crooked or bent
23. Sangir	baŋkuŋ	bent, curved
24. Manobo (Sarangani)	belegkoŋ	crooked, bent
25. Maranao	boyokoŋ	hunchback
26. Iban	cakoŋ	sunken-cheeked
27. Malay	cikoŋ	pit within the collar bone
28. Amis	dəkuŋ	hollowed out; a cleft place in a tree or log
29. Iban	duŋkoŋ	excrescence or bulge, esp. of the head
30. Kayan	ikuŋ	bent, flexed (of the arm or leg)
31. Manggarai	joŋkuŋ	to sit hunched over, crouch
32. Kenyah	kedukoŋ	bent over (of old people)
33. Hawu	keləku	winding (as a river or path)
34. Banggai	keŋkuŋ	wavy, curly (of hair)

(continued)

35. Lamaholot	kəsikū	a bend, curve
36. Kankanaey	kiníkoŋ	tortuous, winding
37. Cebuano	lakún	to loop around, wind into a loop
38. Bikol	likuŋkón	concave, curving inward (like a spoon)
39. Iban	nakoŋ	curving upwards, as the bow or stern of a boat
40. Old Javanese	pələŋkuŋ, pləŋkuŋ	an arch
41. Toba Batak	puhuŋ	clenched (hand); bent, curved
42. Old Javanese	raŋkuŋ	hunched?
43. Itbayaten	salichoŋkoŋ	slightly concave, curving (as a board for making boats)
44. Puyuma (Tamalakaw)	sekuŋ	to string a bow
45. Old Javanese	sidəkuŋ	on bent knee, in a squatting position
46. Isneg	tambilákoŋ	bent
47. Kenyah	tekikoŋ	corner, angle
48. Kankanaey	tibkúŋ	to cause anything to bulge out; swell out (in the center)

*-kuŋ-₂		deep resounding sound (dbl. *-kaŋ-₁, *-keŋ-₂, *-kiŋ)
1. PAN	*kaluŋkuŋ	deep resounding sound
2. PMP	*kuŋ	sound of cooing or barking
3. PAN	*ku(ŋ)kuŋ	deep resounding sound
4. PAN	*qekuŋ	owl
5. PWMP	*sa(ŋ)kuŋ	resounding sound
6. PAN	*tekuŋ	to rap, knock, make a booming noise
7. Javanese	baŋkoŋ	large old frog
8. Ifugaw (Batad)	biʔuŋ	Jew's harp
9. Malay	cəŋkoŋ	to growl; yell or deep bay (of dog)
10. Malay	dəŋkoŋ	the sound of heavy hammering; deep bay of a dog
11. Malay	ləŋkoŋ	to boom out, as a gong
12. Iban	paŋkoŋ	to strike, beat, hammer, usually with ringing sound
13. Iban	raŋkoŋ	loud sound, booming
14. Iban	rəŋkoŋ	noisily, loud sounding (as of s.o. beating on the buttress of a tree)
15. Malagasy	róhona	a sound, as of thunder
16. Amis	takoŋkoŋ	to beat a drum in mourning
17. Malay	taŋkoŋ	a musical instrument of bamboo
18. Javanese	ukuŋ	cooing of a turtledove

*-kup		to enclose, cover (dbl. *-keb$_{-1}$, *-kub$_{-1}$, *-kep$_{-1}$, *-kum)
1. PWMP	*ca(ŋ)kup	to close, enclose
2. PAN	*cukup	to close, cover, fold up
3. PWMP	*e(ŋ)kup	to enclose, cover
4. PWMP	*kupkup	to cover with a thin layer
5. PAN	*lakup	cover, wrapping
6. PWMP	*li(ŋ)kup	to cover, enclose
7. PMP	*ra(ŋ)kup	to gather in cupped hands
8. PWMP	*takup	to cover, enclose
9. PWMP	*tikup	to shut, close, enclose
10. PWMP	*tu(ŋ)kup	to cover
11. PMP	*ukup	to brood, sit on eggs; to cover completely
12. Javanese	aŋkup	calyx (Pigeaud 1938)
13. Ilokano	arákup	to embrace
14. Itbayaten	axokop	sheath, peel of reeds
15. Malay	cəkup	making a scoop of the hand and putting it down on anything; e.g. to catch a fly without killing it
16. Balinese	cikup	to enclose, hold together
17. Lun Dayeh	gerugkup	to fold both arms in front of the body when cold
18. Hiligaynon	hákup	handful, fistful
19. Javanese	iŋkup	closed up, folded
20. Pazeh	kaukup	to brood eggs; to cover chicks with its wings (of a hen)
21. Iban	luŋkup	(of boats, bowls, etc.) turn upside down
22. Iban	paŋkup	squeeze trap for squirrels and rats
23. Samoan	ʔaʔuf-ia	included, embraced, covered
24. Iban	ruŋkup	to invert, turn upside down on, cover completely
25. Bikol	salíkop	to encircle
26. Dumagat (Casiguran)	salokóp	to cup s.t. over s.t. else (as to put a hat on the head, a tin can upside down over a post or stick, etc.)
27. Balinese	saŋkop	to scoop up water with a hand
28. Balinese	səlaŋkup	snare for catching jungle-fowl
29. Iban	suŋkup	miniature house erected over a grave
30. Tiruray	talikuf	to encircle
31. Iban	təkup	to slam shut; to shut up or in
32. Javanese	tlakup	covering leaf of a flower or bud

Note: PWMP *kupkup 'to cover with a thin layer' may be identical to Dempwolff's (1938) *kupkup 'to grasp, hold'), and the forms in Pazeh, Iban and Malay may reflect the doublet *-kub.

*-kuq		bend, curve (dbl. *-kuC, *-kug, *-kuk$_{-1}$, *-kul$_{-1}$, *-kuŋ$_{-1}$, *-kux, *-luŋ$_{-1}$, *-luk/luquk, *-tuŋ$_{-2}$)
1. PMP	*ba(ŋ)kuq	curved, bent
2. PMP	*beŋkuq	bend, curve
3. PMP	*bikuq	bend, curve
4. PMP	*bu(ŋ)kuq	to bend; bent, bowed
5. PMP	*dukuq	to bend over, stoop
6. PWMP	*hukuq	hunched, bent over
7. PAN	*lekuq	go bend; bending part, joint
8. PMP	*likuq	zigzag; winding or curving
9. PWMP	*lukuq	to bend
10. PMP	*pekuq	bend, curve
11. PAN	*pikuq	to bend, curve; bent, curved
12. PWMP	*rekuq	bent, curved; to bend
13. PWMP	*ru(ŋ)kuq	bowed, hunched over, as from age
14. PCEMP	*seku(q)	bent at an angle
15. PWMP	*si(ŋ)kuq	to bend, curve
16. PMP	*tekuq	to bend; hook
17. PWMP	*tikuq	to bend, curve
18. Maranao	balikoʔ	zigzag, crooked
19. Binukid	balukuʔ	to be crooked, curved, in a bent over position
20. Maranao	borəŋkoʔ	bent, crooked; bend
21. WBM	bulegkuʔ	to bend s.t.; to bend, as a trail bends
22. Iban	cakoh	bent, stooping
23. Iban	cukoh	bent forward
24. Old Javanese	ḍəkuh	to bow (respectfully)
25. Cebuano	gukúʔ	bent, bowed, crouched
26. Yakan	kekoʔ	a curve (as in a road)
27. Maranao	latkoʔ	bend
28. BM	leŋkuʔ	bend, curved
29. Iban	paŋkoh	bent, stooping, round-shouldered
30. Hiligaynon	púŋkuʔ	to sit down
31. Kelabit	tekeuʔ	to squat
32. Tagalog	ʔukóʔ	with head bowed low in decrepitude

*-kur		to coo; turtledove	(dbl. *-duR, *-gur, *-ŋur)
1. PWMP	*bekur	to coo; turtledove	
2. Sasak	bəŋkukur	wild dove	
3. Mandar	bukkur	turtledove	
4. Malay	dəŋkur	snore	
5. Manggarai	dukur	turtledove	
6. DPB	əndukur	turtledove	
7. Tetun	kaku	an owl whose voice imitates this word	

(continued)

8. Ida'an Begak	kurkur	a dove
9. Lun Dayeh	sukur	the Spotted Neck Wild Dove
10. Sangir	tarakuku	turtledove
11. Malay	təkukur	turtledove
12. Manggarai	tekur	turtledove
13. Dampelas	uŋkur	to grunt

Note: Manggarai *tekur* could be cognate with Malay *təkukur*, with haplology.

*-kus	to wind around; bundle (dbl. *-kes, *-kis.₂)	
1. PWMP	*buŋkus	bundle, package; to wrap up
2. PPH	*likus	to coil, wind around
3. PWMP	*tuŋkus	bundle
4. Maranao	bakokos	to wrap
5. WBM	bakus	to bind or tie up a person or animal
6. Binukid	balukus	to coil up
7. WBM	burukus	to coil s.t. up, as a rope or a length of rattan
8. Maranao	kokos	to cover, wrap
9. Kankanaey	kusíkus	roll up
10. Ibaloy	pikos	to be coiled, wound up (as a snake)
11. Ilokano	rákus	to bind, make fast; tie with a band
12. Sasak	riŋkus	to bind, tie
13. Balinese	uŋkus	to wrap in a cloth

*-kux	bend, curve (dbl. *-kuC, *-kug, *-kuk.₁, *-kul.₁, *-kuŋ.₁, *-kuq, *-luŋ.₁, *-luk/luquk, *-tuŋ.₂)	
1. PMP	*biŋkuh	to coil, curve, bend, fold
2. PMP	*dekuh	bent, curved
3. PAN	*lekux	to bend, fold; folding part of the body; to curl up
4. PMP	*likuh	winding, curving
5. POC	*noku	to bend, fold
6. PAN	*pikux	bend, curve
7. PAN	*sikux	elbow
8. Ikokano	bakkúko	curved, arched, bent
9. Kapampangan	balikúku	curved
10. Malay	bəliku	a sharp twist or bend in a river
11. Malay	biku	scallop or vandyke border
12. Bare'e	buruku	bent, not upright in walking
13. Rotinese	diku	to fold
14. Buruese	fiku-h	elbow

(continued)

15. Balinese	jəŋku	knee; to kneel
16. Malay	kiku	a turn of the road
17. Erai	kliku	crooked
18. Amis	lopiko	a bend, curve in the road
19. Cebuano	máŋku	s.o. with an arm twisted at the elbow
20. Maranao	talipko	a bend, curve; crooked
21. Malay	tələku	to rest the elbows on any surface
22. Hanunóo	siŋku	right-angle turn in a path, trail or road
23. Kayan	tiko?	a direct route, a devious path

The l-roots

*-laC		to shine; flickering or flashing light (dbl. *-lak$_{-1}$, *-laŋ$_{-1}$, *-lap, *-law)
1. PMP	*bilat	lightning
2. PWMP	*kid(e)lat	lightning
3. PAN	*kilaC	lightning
4. PMP	*silat	to shine
5. Mandar	dala?	lightning
6. Amis	fəlat	to flicker; light appearing and then disappearing, as of a lighted match
7. Label	hil	lightning; to flash, of lightning
8. Manggarai	ilat	to shine, flash, sparkle
9. Erai	jilat	lightning
10. Melanau (Mukah)	kəkəlat	lightning
11. Yakan	lalat	lightning
12. Puyuma (Tamalakaw)	meLaTiLaT	to sparkle, glitter
13. Tetun	milat	polished, bright, glistening

*-laj		to spread out to dry in the sun
1. PEMP	*belaj	to spread out to dry in the sun
2. PMP	*bilaj	to spread out to dry in the sun
3. PPH	*lajlaj	to spread out, as a net or clothes on the ground to dry
4. Tagbanwa (Central)	amlad	to dry s.t. in the sun
5. Keley-i	aplag	to spread a blanket or clothing, often done to dry the item
6. Keley-i	biklag	to spread s.t. out or open; to unroll
7. Aklanon	bueád	to dry out in the sunshine

*-lak.₁		to shine (dbl. *-laC, *-laŋ.₁, *-lap, *-law)
1. PMP	*bilak	to flash, of lightning
2. PMP	*cilak	to shine, of heavenly bodies
3. PMP	*gilak	shine, glitter
4. PMP	*silak	a ray, beam of light
5. Aklanon	bárlak	to shine, radiate
6. Sundanese	burilak	to throw off sparks; glitter (of eyes)
7. Cebuano	busílak	to radiate light, especially the sun
8. Hanunóo	butlák	rising, with reference to the sun
9. Karo Batak	cərlak	to shine, gleam
10. Javanese	ilak	clear, light, bright
11. Kambera	jilaku	to glitter, sparkle, flicker
12. Kambera	lalaku	oily, shiny
13. Kambera	lilaku	to gleam, shine, glitter
14. Kambera	lulaku	to shine, glitter
15. Chamorro	maʔlak	brilliant, sparkling
16. Ngaju Dayak	pila-pilak	shiny (used only of gold and silver)

Note: Also PMP *sirak 'outpouring of light'.

*-lak.₂		to split (dbl. *-laq)
1. PMP	*belak	to crack; split open, as the stomach of a fish
2. PWMP	*siqelak	split or cleaved
3. Kelabit	bilak	a piece of split wood, rattan, or bamboo
4. Keley-i	bulak	to split a rock into pieces
5. Lun Dayeh	filak	to split bamboo or rattan
6. Amis	fitlak	to split open, as flowers beginning to bloom
7. Dumagat (Casiguran)	pilák	to split (as a coconut or other fruit, or to split a bamboo pole into strips)
8. Kayan	pulak	free-splitting, of wood that splits easily
9. Toba Batak	tolak	to split, divide

*-laŋ.₁		to glitter, flash (dbl. *-daŋ.₁, *-naŋ, *-raŋ, *-laC, *-lak.₁, *-lap, *-law)
1. PWMP	*gilaŋ	radiance
2. Kankanaey	bilaŋlán	to glitter; glisten; flash (as fire)
3. Kankanaey	dalán	to glare; to blaze; to shine; to glitter
4. Karo Batak	ərlaŋ	glitter, shine (as the eyes of a cat)
5. Lamaholot	kilã	to glitter, sparkle
6. Tausug	lagaŋlaŋ	bright color; (of a color) bright and glaring
7. Dusun (Kadazan)	puhaŋ	to polish, make shiny

(continued)

8. Malay	rəlaŋ	to glitter, flash
9. Hanunóo	sidláŋ	rising, shining, as of the sun
10. Isneg	siláŋ	moonlight; bright, brilliant, luminous
11. Buginese	welaŋ	ray of light

*-laŋ₋₂		intervening space
1. PMP	*qelaŋ	intervening space
2. PWMP	*selaŋ	interval of space or time
3. Agutaynen	elaŋ	a room divider; walls between rooms
4. Karo Batak	kəlaŋ	the space between two objects
5. Tboli	kilaŋ	space between thighs; to step
6. Toba Batak	laŋlaŋ	interval of space or time
7. Bikol	taláŋ	the space between two things
8. Cebuano	uláŋ	to separate, keep apart

*-laŋ₋₃		spotted, striped
1. PWMP	*balaŋ	spotted, striped, multi-colored
2. PMP	*belaŋ	spotted, dappled
3. PWMP	*palaŋ	spotted
4. PWMP	*patelaŋ	striped, in bands of different color
5. Yami (Iranomilek)	be-teŋelaŋ	(of animals) striped, patchy in color
6. Gorontalo	hulaŋo	sickness which leaves striped marks on the calves of the legs
7. Malay	pəlaŋ	striped; banded
8. Kelabit	pilaŋ	striped
9. Pazeh	raŋaraŋ	clouded leopard

*-lap		to flash, sparkle, shine (dbl. *-daŋ₋₁, *-naŋ, -laC, *-lak₋₁, *-laŋ₋₁, *-law)
1. PMP	*dilap	to sparkle, shine
2. PMP	*gelap	lightning that strikes s.t.
3. PMP	*gilap	radiance
4. PMP	*ilap	to flicker, of flames
5. PMP	*kelap	to shine, sparkle, twinkle
6. PWMP	*kid(e)lap	to glitter, flash, sparkle
7. PMP	*kilap	flash, sparkle
8. PPH	*kis(e)láp	to sparkle, glitter
9. PMP	*silap	to sparkle, dazzle

(continued)

10. Ida'an Begak	bəkədilap	to twinkle
11. Old Javanese	burilap	flickering (as a torch)
12. Manggarai	celap	dazzled by blinding light
13. Tagalog	diklap	to sparkle
14. Karo Batak	ərlap	to glitter, as gold or precious stones
15. Old Javanese	gurilap	flickering (as a torch)
16. Old Javanese	hulap	dazzlement, being unable to look at a bright light
17. Aklanon	ídlap	to shine, glow
18. Dumagat (Casiguran)	ʔisláp	blinding reflection (from something shiny)
19. Malagasy	jélaka	clearness, brightness, sudden glitter, as a torch at night
20. Kambera	kalila	brilliance, radiance (as from the sun)
21. Ida'an Begak	bə-kədilap	to twiinkle
22. Malay	kərlap	to flash, glisten, as the wings of flying butterflies
23. Pangasinan	kírlap	to glitter
24. Batak (Palawan)	kudláp	lightning
25. Ifugaw	kuhílap	what is brilliant, glittering, gleaming
26. Kambera	lula	shining, glittering
27. Manggarai	nilap	lightning
28. Dobel	ŋela	lightning
29. Tolai	pala	to flash
30. Malay	rəlap	to glint, flash
31. Aklanon	síglap	to sparkle, shine
32. Old Javanese	sulap	blinding (by a bright light)
33. Lau	tala	to shine, of sun or moon
	talaf-i	to shine on, enlighten, lighten up
34. Wolio	tila	light, radiation, glare, mirror; to shine upon, irradiate, illuminate
35. Cebuano	uyláp	to flare up, burst into flames

Note: Also Ilokano *sírap* 'to dazzle', PMP *kilab* 'to flash, sparkle'.

*-laq-1	to split (dbl. *-lak)	
1. PWMP	*balaq	to split, divide
2. PAN	*belaq	to split
3. PWMP	*bilaq	part split off; chip, splinter
4. PAN	*Celaq	to crack, split; a fissure
5. PMP	*kelaq	crack, split
6. POC	*pwalaq	to split wood
7. PAN	*silaq	to split

(continued)

8. Lau	ala	to split across
9. Buruese	gela-h	cracked
10. Ida'an Begak	ila?	to split; to operate (surgery)
11. Fijian	katela	to split or divide into two parts
12. Ngadha	nela	to split, crack; cracked (as bamboo)
13. Amis	pəla?	to split into two
14. Malay	rəlah	torn along the seams; ripped (of tailored garments)
15. Kelabit	təkila?	chips
16. Bunun	timpusla?	to split

Note: Also PMP *siraq 'to split open', Old Javanese təlā 'cleft, crevice, crack, narrow opening'.

*-laq.$_2$		tongue; to lick
1. PAN	*dilaq	to lick
2. PPH	*tak(e)laq	to click with the tongue
3. PWMP	*zelaq	tongue
4. Itawis	híla	tongue
5. Chamorro	hula?	tongue; sticking out tongue
6. Nias	lela	tongue
7. Tae'	lila	tongue
8. Kayan (Uma Juman)	ñila	to lick

*-lat.$_1$		to open the eyes wide
1. PMP	*bilat	to open the eyes wide
2. PMP	*bulat	to open the eyes wide, stare with round eyes
3. PMP	*hilat	wide open, of the eyes
4. PMP	*kilat	to open the eyes wide
5. PMP	*kulat	to open the eyes
6. Toba Batak	alat	to look around, observe
7. Cebuano	búdlat	bulging eyes; for the eyes to bulge
8. Tagalog	buklát	open, opened
9. Tagalog	dílat	the act of opening of the eyelids
10. Lun Dayeh	kalat	awakening, opening the eyes
11. Yakan	kəllat	to be open (of eyes)
12. Itbayaten	korilat	wide-opened eyes
13. Aklanon	múdlat	to open up the eyes very wide
14. Tagalog	múlat	open (said of eyes)
15. Pazeh	mu-xurat	to open one's eyes

*-lat.₂		scar
1. PMP	*bilat	scar
2. PPH	*bik(e)lat	scar
3. PPH	*pik(e)lat	scar
4. PWMP	*pilat	scar
5. PWMP	*ulat	scar
6. Lau	bala	white from scars, of the face or body
7. Ayta Abellen	dəlat	wound
8. Rembong	gilat	scar
9. Arosi	hura ~ hu-hura	scar
10. Kankanaey	kalálat	scarred
11. Dumagat (Casiguran)	péklat	scab of a sore; to scab over

Note: Also Isneg *inírat* 'the scar left on a tree when a nest of bees has been removed from it'.

*-law		dazzling light (dbl. *-daŋ.₁, *-naŋ, -laC, *-lak.₁, *-laŋ.₁, *-lap)
1. PWMP	*gilaw	gleaming, luminous
2. PWMP	*ginelaw	dazzling light
3. PWMP	*kilaw	luminous, brilliant, full of light
4. PWMP	*kinelaw	dazzling light
5. PMP	*nilaw	bright light
6. PMP	*qilaw	torch; to illuminate
7. PMP	*silaw	dazzling, of light
8. Malay	kəlip-kəlau	shimmer
9. Fijian	mimilo	shiny, glossy
10. Bare'e	pilo	to flicker, of a fire
11. Kelabit	silo	shiny
12. Dumagat (Casiguran)	taŋláw	light, illumination, beam
13. Buginese	wilo	dazzled by glare

*-leb		to sink, disappear under water (dbl. *-led, *-lem, *-lep.₂)
1. PMP	*celeb	to sink, submerge
2. PWMP	*eleb	to disappear under water
3. PMP	*lebleb	to sink, disappear under water
4. PMP	*teleb	to sink, vanish from sight
5. Manggarai	belep	to sink
6. Manggarai	delep	to sink
7. Old Javanese	kələb	to be submerged, sink
8. Manggarai	melep	to sink (< *-lep?)
9. Lau	ōlo	to sink, as a stone sinks in water
10. Manggarai	selep	to sink, submerge
11. Uma	ma-tala	to sink
12. Cebuano	túnlub	stick, dunk s.t. completely or partly into a liquid or solid

*-leC		interval, gap, intervening space
1. PMP	*belet	interval
2. PAN	*qeleC	interval, intervening space
3. PPH	*salét	to insert between, intersperse
4. PMP	*selet	to insert between
5. Aklanon	hueót	room divider, partition, division
6. Bikol	sílot	to crawl through a tight space
7. Maranao	soŋgalət	intervene, interrupt
8. Javanese	wilət	(of gamelan music) interval between notes

*-led		to sink (dbl. *-leb, *-lem, *-lep$_{-2}$)
1. PWMP	*qeled	to sink
2. Bare'e	dolo	to submerge, go under water
3. Dumagat (Casiguran)	ʔəglád	to submerge; to push someone's head under water
4. Maranao	galəd	to sink, as in liquid; immerse
5. Isneg	pálád	to sink, go to the bottom
6. Cebuano	únlud	to sink, submerge

*-lem$_{-1}$		dark; obscure (dbl. *-Cem, *-dem)
1. PMP	*alem	dark; night
2. PAN	*beNelem	overcast, twilight
3. PMP	*delem	dark of the moon, moonless night
4. PPH	*dik(e)lém	dark, obscure, cloudy
5. PAN	*dulem	dark
6. PMP	*elem	shade, darkness
7. PWMP	*gelem	dark, overcast, cloudy
8. PMP	*halem	night; dark
9. PMP	*kalem	dark
10. PAN	*kelem	dark, overcast, visually obscure
11. PWMP	*kulem	dark, dim
12. PAN	*lemlem	dark, of weather; overcast
13. PMP	*malem	night; darkness
14. PMP	*qilem	dim, dark, obscure
15. PWMP	*selem	dark
16. PWMP	*silem	to fade from sight; gloom, obscurity
17. PAN	*Sulem	dimness, twilight
18. PWMP	*tilem	dark, darkened
19. Malay	balam	dimly seen
20. Ngaju Dayak	bilem	black
21. Ayta Abellen	daəm	clouds; cloudy

(continued)

22. Ayta Abellen	dəgləm	evening, become nighttime
23. WBM	dukiləm	to become night
24. Balinese	guləm	cloud, cloudy
25. Agta (Central Cagayan)	hiklem	night
26. Mapun	hilom	a bruise
27. Kanowit	jekelem	dark
28. Cebuano	kilum-kílum	dusk
29. Ilokano	lúlem	overcast, clouded over, darkish
30. Tae'	masullan	dark, of color
31. Kelabit	milem	purple, color of a bruise
32. Singhi	ŋarom	night
33. Kenyah	pelem	blinking, blurred vision, dim
34. Pazeh	ma-rarem	cloudy; to have poor vision
35. Maranao	saləm	darkening, as of the sun
36. Paiwan	selem	the dark, the World of the Dead
37. Bikol	sulóm	dark, obscure, as a house without lights
38. Kavalan	telem	dark
39. Old Javanese	uləm	shining (only) palely, dim, wan, languishing, faded
40. Paiwan	velʸelem	cloud shadows; overcast
41. Pazeh	ma-xurum	cloudy
42. Itbayaten	yosləm	dark-complexioned

Note: Also Casiguran Dumagat *kulimlím* 'for the sun to be darkened by a thin overcast (but with a slight amount of sunlight coming through)'.

*-lem.₂		in, inside; deep
1. PAN	*dalem	in, inside; deep
2. PWMP	*ha-dalem	deep
3. Kayan	alem	deep
4. Lampung	delom	inside
5. Kayan (Uma Juman)	haləm	in, inside
6. Kapampangan	lálam	the underneath area; depth
7. Favorlang	lallum	inside
8. Kenyah (Long Wat)	liləm	deep
9. Tboli	nelem ~ ŋelem	deep, as a hole; far down
10. Murik	tələm	in, inside

Note: Many Austronesian languages reflect *dalem both in the meaning 'in, inside', and the meaning 'deep'. However, many languages in the Philippines and Sabah distinguish these meanings with reflexes of *dalem meaning 'in, inside', and reflexes of *ha-dalem meaning 'deep'. I am indebted to David Zorc for pointing out the evidence for PAN *Sa-, PMP *ha- 'prefix of measurement'. For an argument as to why 'inside' and 'deep' came to be associated with the same morpheme cf. Blust (1997).

*-lem.₃		to sink, submerge (dbl. *-leb, *-led, *-lep.₂)
1. PWMP	*salem	to dive, immerse oneself
2. PAN	*selem	to sink, immerse oneself in water
3. PMP	*telem	to sink, disappear under water
4. PWMP	*tilem	to dive, submerge
5. Manggarai	delem	to sink
6. Balinese	kələm	to sink, go under, drown
7. Javanese	siləm	under water; to become immersed
8. Malay	təŋəlam	sinking, being submerged

*-len		to swallow
1. PCEMP	*belen	to swallow whole, without chewing
2. PMP	*telen	to swallow
3. PMP	*tilen	to swallow
4. Bikol	hálon	to swallow
5. Rotinese	sila, sile	to swallow
6. Iban	təŋkəlan	to choke, have food stuck in the throat, or 'go the wrong way'

*-leŋ		sound that stuns a person
1. PWMP	*tileŋ	piercing sound
2. Malay	biŋoŋ alaŋ	knocked sily; bewildered
3. Malay	gəlaŋ	clang, ringing sound
4. Maranao	olalaŋ	echo
5. Ilokano	pellén	to be stunned, rendered senseless (by a blow on the head, the report of a gun, the crash of thunder, etc.)
6. Ilokano	síleŋ	what annoys the ear through its strong sound (drums, bells, etc.)

*-lep.₁		shiny, bright (dbl. *-lap)
1. PMP	*dilep	to flash, sparkle, shine
2. Dumagat (Casiguran)	ʔiláp	to reflect (of something shiny)
3. Lun Dayeh	kilep	lighting; blinking
4. Manggarai	mbilep	to shine
5. Banggai	ŋgilep	to glitter, shine, gleam
6. Manggarai	silep	to shine, glitter

*-lep₂		to sink, submerge (dbl. *-leb, *-led, *-lem₋₃)
1. PWMP	*delep	to immerse, submerge
2. PMP	*le(p)lep	to submerge, sink
3. PWMP	*selep	to enter, penetrate
4. Maranao	aləp-aləp	to sink deeply, penetrate throughout
5. Javanese	ilep	to put under water, soak
6. Maranao	oləp	to sink deeper
7. Javanese	iləp	submerged in water

*-leR		defective vision
1. PMP	*bileR	cataract of the eye
2. Molbog	bolog	blind
3. Kapampangan	bulág	blind
4. Maranao	iləg	cross-eyed, wall-eyed

*-liŋ₋₁		a clear ringing sound
1. PMP	*eliŋ	high-pitched sound
2. PPH	*kililíŋ	ringing of a bell
3. PPH	*kilíŋ	ringing of a bell
4. PPH	*kuliliŋ	ringing of a small bell
5. PMP	*liŋ	the sound of ringing
6. PWMP	*suliŋ	bamboo flute
7. PWMP	*tiliŋ	high-pitched ringing sound
8. Karo Batak	aloliŋ	echo
9. Bare'e	dili	onom. for a tap on a drum, or against metal
10. Toba Batak	huliŋ	to sound out
11. Javanese	jléŋ	metal clanking
12. Hanunóo	káliŋ	to jingle
13. Puyuma (Tamalakaw)	kamriŋ	k.o. middle-sized bell
14. Old Javanese	kəléŋ	to ring a prayer-bell
15. Sangir	ləlliŋ	echo
16. Sangir	liliŋ	faint sound from the distance
17. Ngadha	ruli	to resound, ring
18. Amis	takliŋ	a bell musical instrument tied around the waist and dangling from the side or back
19. Karo Batak	uliliŋ	echo

*-liŋ₋₂		cross-eyed
1. PPH	*paq(e)liŋ	visible defect of the eyes
2. PWMP	*zeliŋ	cross-eyed; to squint

(continued)

3. PWMP	*zuliŋ	cross-eyed
4. Bimanese	giri	cross-eyed, squinting
5. Bontok	kosliŋ	to squint, as in bright sunlight
6. Iban	səliŋ	to look from the corner of the eye, glance
7. WBM	siliŋ	cross-eyed

Note: Also Maranao *pilaŋ* 'poor of sight, blind', Old Javanese *dilaŋ* 'cross-eyed'.

*-liŋ$_{-3}$		to tilt to the side
1. PMP	*kiliŋ	leaning sideways, listing (as a boat)
2. PMP	*liliŋ	askew, in a slanting direction
3. PWMP	*tiliŋ	tilted, at an angle
4. Bikol	kulíŋ	to turn the face away
5. Malay	səŋkəliŋ	having a leg or arm on the other at a slight angle
6. Bikol	talíliŋ	to tilt the head to one side

Note: Apparently distinct from both *-liŋ$_{-2}$ 'cross-eyed', and *-liŋ$_{-4}$ 'to turn, revolve, turn around'.

*-liŋ$_{-4}$		to turn, revolve, turn around
1. PWMP	*baliliŋ	to turn, revolve
2. PWMP	*baliŋ	bent, twisted
3. PMP	*biliŋ	to turn, revolve
4. PWMP	*galiŋ	to roll, grind
5. PWMP	*geliŋ	to roll over and over, as a stone
6. PMP	*giliŋ	to grind, as by rolling over
7. PWMP	*guliŋ	to roll
8. PWMP	*iliŋ	to shake the head in negation, shake the head from side to side
9. PMP	*liliŋ	to go in a circle, circle around s.t.
10. PMP	*paliŋ	to turn (as the prow of a boat)
11. PMP	*puliŋ	to turn round, rotate
12. Banggai	aliliŋ	to turn around, go back
13. BM	boliŋ	to turn
14. Mandar	guliliŋ	to wander this way and that
15. Kenyah	kəliliŋ	going round and round; circular
16. Bikol	pilíŋ-pilíŋ	to shake the head from side to side, indicating negation
17. Manggarai	tiliŋ	to turn, revolve
18. Uma	wulili	whirlpool

Note: Also Bintulu *bekeleŋ* 'to turn around and return to where one started', Kenyah *mileŋ* 'to shake the head'.

*-lip	to insert, slip in	
1. Aklanon	húlip	to insert into
2. Maranao	lilip	to insert, squeeze in or through
3. Karo Batak	salip	to insert between
4. Malay	səlip	jammed in between two surfaces; wedged in

*-liq	to return, restore	
1. PWMP	*puliq	to return to its previous conditon
2. PMP	*suliq	to reverse, turn around
3. PMP	*uliq	to return home; return s.t.; restore; repair
4. Itbayaten	havili	getting back, taking back, restoring original status or position
5. Bontok	sobli	to return
6. Itbayaten	vili	returning, restoring, going around

*-liR	to flow (dbl. *-luR)	
1. PMP	*aliR	to flow
2. PWMP	*qiliR	to flow downstream
3. PMP	*saliR	to flow
4. Bare'e	gili	to stream, flow down
5. Malay	jalir	to flow
6. Sangir	sellihe?	current
7. Tetun	suli	to flow

*-lit.₁	to caulk; adhesive material	
1. PMP	*belit	viscous, sticky
2. PMP	*bulit	glue, paste; to stick, caulk
3. PMP	*dalit	glue, paste, plaster; to caulk
4. PWMP	*dulit	to stick, adhere
5. PMP	*pilit	to paste, stick to
6. PMP	*pulit	glue, paste; stick, to caulk
7. Kankanaey	dumlít	to cement, putty with; to stick, adhere
8. Kayan	halit	stuck or pasted on
9. Aklanon	haplít	to paste, attach, put up
10. Kayan	jepalit	sticky, soft, plastic (as clay)
11. Kayan	kalit	clay
12. Javanese	kəlét	firmly attached
13. Sasak	malit	kind of clay used to make blowpipe pellets

(continued)

14. Tetun	nulit	sticky
15. Keley-i	paklit	to throw mud or a similar sticky object at s.t. or s.o.
16. Malay	palit	smudging
17. Keley-i	pudlit	to plug or cover holes or cracks
18. Ida'an Begak	səllit	sticky

Note: Also Puyuma *salit* 'plaster made of clay mixed with straw and cow dung', *sapəlit* 'sticky and dirty with sweat', both with *l* < *N.

*-lit.₂	difficult, complicated	
1. PWMP	*salit	difficult, hard
2. PWMP	*sulit	difficult, complicated
3. Iban	alit	puzzled, dumbfounded
4. Cebuano	búlit	burden with problems, debt, etc.

Note: Possibly modelled on *-lit 'to wind, twist; be intertwined'.

*-lit.₃	to wind, twist; be intertwined (dbl. *-lut)	
1. PMP	*belit	to twist around
2. PMP	*bilit	to intertwine (as strands in making rope)
3. PAN	*litlit	to wind around, bind by winding around
4. PAN	*palit	circling of a rope (in tying)
5. PMP	*pulit	to twine around
6. Toba Batak	alit	wound, turned, twisted
7. Javanese	bəṇḍalit	to twist, (inter)twine
8. Bare'e	gili	to wind, twist around
9. Kayan	kelilit	a bandage; to bandage
10. Kankanaey	litílit	to twist, wring
11. Fijian	qali	to twist together, roll together, as in making string
12. Fijian	tali	to plait, interweave, twist

*-luk (or *-luquk?)	bend, curve (dbl. *-kuk.₁)	
1. PMP	*beluk	to bend, curve
2. PWMP	*biluk	to wind, curve, turn to the side; to tack, sail into the wind
3. PWMP	*celuk	curved area; corner, angle
4. PWMP	*keluk	bend, curve
5. PCEMP	*kiluk	wavy or zigzag line

(continued)

6. PWMP	*luquk	bay
7. PMP	*peluk	to bend, curve
8. PAN	*piluk	crippled; bent or twisted, of the leg or foot
9. PWMP	*qeluk	bend, curve
10. PWMP	*seluk	curved area, angle
11. PWMP	*siluk	corner; curve
12. PWMP	*teluk	curved, as a shoreline
13. Itbayaten	ahxok	curving (of wire)
14. Fijian	belu	to bend, curve, of hard things, e.g. tin
15. Javanese	baŋkəluk	bent at the end (as a pen with a bent point)
16. Sangir	biruluʔ	to wind into coils (as a snake)
17. BM	bolotuk	curved bamboo lath inserted in the ground
18. Sundanese	dəluk	somewhat bent in form (as very tall persons)
19. Rotinese	dilu	to bend over, turn down, bend down
20. Ngaju Dayak	hulok	bay, gulf
21. Rejang	iloʔ	tortuous, winding
22. Malay	jəlok	pronounced concavity
23. Erai	klulu	knee
24. Niue	pelu	to fold, bend; an ancient type of curved club
25. Dumagat (Casiguran)	súluk	corner
26. Kambera	tambàluku	bent
27. Bare'e	tiŋkolu ~ tiŋkoyu	to curl up, as a snake; the curved part of something (hoop, strand, loop)
28. Bare'e	walu	to wind around

*-luŋ$_{-1}$		bend, curve
1. PWMP	*beluŋ	bend, curve
2. PWMP	*biŋkeluŋ	to curve, bend around
3. PMP	*eluŋ	bend, curve
4. PWMP	*galuŋ	curve
5. PWMP	*guluŋ	to roll up
6. PWMP	*kaluŋ	curved
7. PWMP	*keluŋ	coil, curl, undulation
8. PAN	*kiluŋ	curved, bay
9. PMP	*kuluŋ	to curl, curve
10. PPH	*paluŋ	wave at sea, sea swell
11. PMP	*zuluŋ	prow
12. BM	beluŋ	dented, bent, curved

(continued)

13. Iban	gəloŋ	coil
14. Lamaholot	kəlulũ	spool
15. Nggela	malu	to droop, as a branch in the heat; bow the head
16. Javanese	məluŋ	to bend, bow
17. Javanese	pəntəluŋ	bending, bowed (of a branch)
18. Itbayaten	piloŋ	wavy, of rough sea
19. Iban	səŋkiloŋ	coil
20. Iban	səŋkuloŋ	coil, skein

*-luŋ.₂		to shelter, shade (dbl. *-duŋ)
1. PMP	*aluŋ	shade, shadow
2. PWMP	*heluŋ	shade, shelter
3. PWMP	*keluŋ	shield
4. PPH	*luŋáluŋ	shelter, temporary hut
5. PS	*səluŋ	to shelter (from sun or rain)
6. PPH	*siluŋ	shelter, space under s.t.
7. PMP	*suluŋ	to shelter, seek shelter
8. Tiruray	anluŋ	a shadow
9. Maranao	daloŋ	to hide behind, shelter
10. Maranao	diloŋ	to hide behind a cover or protection
11. Tiruray	galuŋ	shade o.s.; in the shade
12. Dumagat (Casiguran)	ʔilóŋ	covered, sheltered (of cultivated crops)
13. Tagbanwa (Central)	inloŋ	shade
14. Soboyo	kamaloŋ	shadow, shade
15. BM	luluŋ	what serves to give shade
16. Tongan	malu	shaded or sheltered; safe, secure
17. Sangir	paluŋ	umbrella, parasol
18. Kayan	sekaluŋ	to carry protectively
19. Bunun	taluluŋ	conical field hat
20. Tagbanwa (Kalamian)	uluŋ	shadow

*-luR		to flow (dbl. *-liR)
1. PWMP	*aluR	pond, stream
2. PAN	*iluR	river channel
3. PMP	*iluR	spittle, saliva
4. PWMP	*saluR	pond, stream
5. PWMP	*suluR	flowing water; flood
6. PWMP	*uluR	to flow, stream together
7. Melanau (Matu)	beloh	mild flood

(continued)

8. Palauan	beríus	current (in ocean, river, etc.)
9. Berawan (Long Terawan)	jullu	to flow
10. Rejang	palua	irrigation channel
11. Melanau (Matu)	suloh	serious flood
12. Wolio	wilu	saliva, spittle; to slaver

*-lus		to slip off, slide down (dbl. *-rus)
1. PMP	*luslus	hernia; to slip down, slip off
2. PAN	*pelus	to slip off
3. PWMP	*ulus	to slide down
4. Itbayaten	aʔxos	molting
5. Malay	bəlus	loose, slipping freely off and on (of a ring)
6. Tagalog	bululós	diarrhoea
7. Dumagat (Casiguran)	búlus	diarrhoea
8. Malay	cəlus	loose-fitting; fitting comfortably or easily, of a ring that slips on or off
9. Hiligaynon	dáplus	to slide off
10. Tiruray	dilus	(of a surface) water-repellant
11. Dusun (Kadazan)	gahus	to loosen and slide off
12. Tboli	glus	to slip out of one's hand or grip
13. Mapun	haplus	to slide back and forth or up and down
14. Hawu	hilu	snake skin
15. Cebuano	hilús	for s.t. tied securely in place to slip off, move out of place by sliding
16. Tboli	lus	a shed skin, as of snake, chicken legs or newborn baby
17. Samoan	mamulu	to slip out, as a knife from one's hand
18. Tboli	melus	to loosen, come off
19. Manggarai	ŋolus	loose, slipping freely off and on
20. Hiligaynon	palús	to slip off, free from grip
21. Amis	pololos	to slide down out of position
22. Aklanon	pudlos	to get free, slip out/away, escape
23. Maranao	tamlos	fishhook that loses fish easily
24. Malay	təlus	slipping through without difficulty; fitting loosely, of a ring fitting comfortably on a finger, etc.
25. Bare'e	wulu	slipped off, pushed off

*-lut		to twist around (dbl. *-lit.₃)
1. PWMP	*balut	to roll or wrap s.t. up
2. Tagalog	bílot	small roll or pack
3. DPB	bulut	tangled, twisted
4. Sasak	gəlut	to wrestle
5. Balinese	ilut	to twist, turn

The m-roots

*-mak		mat
1. PWMP	*amak	mat
2. PWMP	*hamak	mat
3. PMP	*lamak	mat
4. Dumagat (Casiguran)	makmak	leaves which are spread out on the ground to sit on or sleep on
5. Sundanese	samak	pandanus mat

Note: Also Agutaynen *amek* 'woven mat used for sleeping, drying rice, etc.'.

*-mek		to crush, pulverize; powder (dbl. *-bek.₂, *-dek, *-muk.₁, *pek)
1. PAN	*Cumek	to pulverize; crumble
2. PR	*lameke	rotten (as a log)
3. PMP	*mekmek	broken to bits
4. PMP	*remek	to crush, pulverize; crumble
5. PMP	*Remek	to crush, pulverize; crumble
6. PPH	*Rumek	to crush, smash
7. Pazeh	demek	rotten (as a log)
8. Hiligaynon	dúgmuk	to crush, break by pressure
9. Bontok	lagmə́k	to break into pieces, as a reed; crush, as a dried reed or bamboo
10. Chamorro	lommok	to pound, beat, pulverize, make into pulp by beating
11. Kankanaey	mayəkmə́k	fine, pulverized (< *mekmek?)
12. Waray-Waray	rugmók	fallen, collapsed, crushed
13. Ilokano	simék	to pulverize, triturate, comminute
14. Paiwan	sulʸamek	powder
15. Pangasinan	təmə́k	to crush
16. Itbayaten	xoʔmək	fine particles, esp. of corn

*-mes		to squeeze, knead
1. PMP	*kemes	to grip, compress, squeeze
2. PAN	*mesmes	to grasp, grip, squeeze
3. PAN	*Rames	to squeeze, knead, mix with the hands
4. Bikol	gumós	to squeeze with the hand
5. Rejang	namɨs	to squeeze coconut juice from the pith
6. Amis	ʔməc, ʔiməc	to grab s.o. at the neck with hands to choke
7. Kalagan	tigmɨs	to squeeze in the hand

Note: Also Casiguran Dumagat *lámas* 'to squeeze s.t. with the hands, bunching it together, as in the working of dough in making bread'.

*-mis		sweet
1. PMP	*emis	sweet taste
2. PAN	*ma-hemis	sweet
3. PMP	*taq(e)mis	sweet
4. Ibaloy	amis	sweetness, delicious taste
5. Buli	mismis	sweet
6. Palawano	regmis	sweetness, sweet or sugary taste
7. Bintulu	təmis	sweet
8. Balangaw	umís	sweet

*-mit		small, slight
1. Old Javanese	dəmit	small, fine, thin, slender
2. Ida'an Begak	lumit	fine
3. Iban	mimit	a little, few; slightly
4. Iban	mit	small, little
5. Kayan	tumit	not reaching to, too short

*-muk.$_1$		to crush, pulverize; powder
1. PWMP	*mukmuk	crumbs that drop when eating rice
2. PMP	*remuk	to crumble, fall to bits
3. PWMP	*Remuk	crumbs
4. Bontok	gomók	to crush; crumble, as dry food or soft
5. Bisaya (Limbang)	ramuk	rotten (wood)
6. Ilokano	timók	powder, dust; fine particles to which s.t. is reduced by grinding, etc.

Note: Also Iban *rumoh* 'shatter, (of food) crumble'.

*-muk₂		mosquito, sandfly
1. PMP	*emuk	mosquito
2. PMP	*lamuk	mosquito
3. PMP	*ñamuk	mosquito
4. Ibaloy	imok	general term for several small, blood-sucking flying insects
5. Kenyah	jamok	sandfly
6. Tae'	katamuk	small mosquito
7. Uma	koromuʔ	mosquito
8. Manggarai	lemuk	sandfly, small biting mosquito
9. Ma'anyan	mamuʔ	mosquito
10. Ibanag	ləmmuʔ	mosquito
11. Ibanag	nəmmuʔ	mosquito
12. Mono-Alu	simuʔu	midge

*-mul		to hold s.t. in the mouth, suck on s.t.
1. PPH	*amúl	to suck on s.t.
2. PAN	*mulmul	to hold in the mouth and suck
3. Kapampangan	akmúl	to swallow s.t.
4. Bikol	hamól	to suck with the lips
5. Ibaloy	kamol	to put s.t. into the mouth and then do nothing with it, not chew, swallow or spit out (as candy, thermometer)
6. Maranao	kimol	movement of the mouth in eating
7. Seediq	umul	to suck on s.t. held in the mouth

Note: Also Ayta Abellen *aməl-aməl* 'to suck on, as candy', Casiguran Dumagat *ʔəmél* 'to suck on s.t., as candy', *məlməl* 'to stuff food into the mouth'.

*-muR₁		dew
1. PWMP	*dahemuR	dew
2. PPH	*hamuR	dew
3. PMP	*lamuR	dew
4. PAN	*ñamuR	dew
5. Cebuano	yámug	dew

*-muR₂		to gargle, rinse the mouth
1. PMP	*emuR	to hold in the mouth
2. PMP	*kemuR	to hold liquid in the mouth; to gargle
3. PMP	*kumuR	to gargle, rinse the mouth
4. PAN	*muRmuR	to gargle, rinse the mouth

(continued)

5. PAN	*qumuR	to fill the mouth with food or water
6. PAN	*timuRmuR	to rinse the mouth
7. PAN	*umuR	to hold in the mouth
8. Lamaholot	gəməmu	to gargle
9. Hiligaynon	límug	to gargle
10. Manggarai	memur	to rinse the mouth
11. Ilokano	mulúmog	to rinse the mouth, gargle

The n-roots

*-nab		to flood, inundate (dbl. *-ñeb, *-ñej, *-ñep)
1. PMP	*lanab	high water, flood
2. PAN	*lenab	to be flooded
3. Keley-i	hiknab	to be covered with water, flooded
4. Ilokano	nabánab	flooded, inundated
5. BM	mo-yanap	overflow (of river)

*-naŋ		to shine, sparkle (dbl. *-daŋ$_{-1}$, *-laŋ, *-raŋ)
1. PWMP	*linaŋ	shiny, shining
2. PWMP	*sinaŋ	to shine; radiance
3. Sundanese	baranaŋ	to shine, to twinkle (of many lights)
4. Keley-i	bənaŋ	ray of light; light of sun, moon, stars, or lamp
5. Tagalog	kináŋ	shininess, lustre
6. Tiruray	səlinaŋ	direct sunlight
7. Maranao	sənaŋ	sun
8. Maranao	somarinaŋ	to shine
9. Maranao	tənaŋ	to shine; lustre

*-naw		lake, pond; enclosed body of water
1. PAN	*banaw	lake, pond
2. PAN	*danaw	lake
3. Dusun (Kadazan)	anaw	paddyfield, wet ricefield
4. Cebuano	bagánaw	pool of water (after rain, etc.)
5. Binukid	balanʔaw	puddle (of water)
6. WBM	belenew	puddle; surface water; "run off" after a rain
7. Ilokano	danʔáw (+M)	lake, pond

(continued)

8. Tagalog	lánaw	pool, small lake; lagoon
9. Amis	nawnaw	to be swamped, surrounded by water
10. Keley-i	tənəw	for a small amount of liquid to accumulate in a container, or in a small area

*-nek		sound, of sleep
1. Agutaynen	anək	deep sleep; sound sleep
2. Malay	jənak	deep or sound, of sleep
3. DPB	lénək	to sleep soundly
4. Bikol	nánok	sound, deep (sleep)
5. Maranao	nək	sound asleep
6. Hanunóo	ranúk	sleep

Note: Also Malay *ñəñak* 'deep, of sleep'.

*-nem		cloud, cloudy
1. PWMP	*kunem	cloud
2. Banggai	bunom	cloud
3. Bolinao	gúnɨm	cloud
4. Kavalan	ranem	cloud (white or black), fog

Note: Also Agutaynen *onom* 'cloud'.

*-neŋ		still, tranquil, calm, as water
1. PMP	*qeneŋ	quiet, still, at rest
2. PWMP	*teneŋ	calm, still, as the surface of water
3. Dumagat (Casiguran)	gánəŋ	to remain motionless (of fish in the water)
4. Itbayaten	hinaknəŋ	letting liquid stand still
5. Keley-i	ʔinəŋ	quietness, calmness, stillness
6. Ayta Abellen	kənəŋ	stagnant, still water; no current
7. Kayan	leneŋ	the calmness of still water in river or pond
8. Hanunóo	linuŋ	peace, quiet, calmness, tranquillity
9. Kavalan	m-rikneŋ	to stand still, motionless
10. Bikol	ma-tunínoŋ	calm, peaceful, quiet, still, tranquil
11. Toba Batak	unoŋ	still, calm (of water)

*-neR		to hear
1. PWMP	*teneR	voice
2. Subanen	bonug	to hear
3. Aklanon	dánog	to echo; to sound, make a sound
4. WBM	dinəg	to hear
5. Aborlan Tagbanwa	geneg	to hear, listen
6. Maranao	kanəg	sense of hearing; hear
7. Tagalog	kiníg	to listen to (someone, or a sound)
8. Cebuano	lánug	loud, resonant; echo
9. Bikol	lusnóg	deaf (said only in anger)
10. Maranao	nəg	to hear; hearing
11. WBM	peminəg	to listen to something; to obey

Note: Also Ayta Abellen *tonoy* 'sound or noise'.

*-ni		to hide, conceal oneself or s.t.
1. PWMP	*buni	invisible nature spirit
2. POC	*mumuni	to hide
3. PWMP	*sambuni	to hide; hidden
4. PWMP	*tabuni	to hide; hidden

*-niŋ		clear, limpid (water)
1. PMP	*niŋniŋ	clear, of water
2. Javanese	bəniŋ	clear (of liquid, eye, sound)
3. Javanese	(ə)niŋ	clear (of insight)
4. Ngaju Dayak	giniŋ	to shine, sparkle
5. Hiligaynon	híniŋ	shiny, polished
6. Dusun (Kadazan)	huniŋ	varnish

*-nit		to remove, detach (dbl. *-kit$_{-3}$)
1. Hanunóo	ánit	molting of animals
2. Mansaka	bignit	to remove
3. Keley-i	buʔnit	to remove a peeling or skin, as of betel nut
4. Aklanon	húgnit	to snatch, seize, grab
5. Agutaynen	init	to remove lice nits or eggs from hair
6. Agutaynen	kanit	to remove hair from the hide of a pig which has been butchered
7. Ibaloy	Kaʔnit	to remove the meat from a bone
8. Mansaka	lanit	to remove (as a piece of paper stuck on the wall)
9. Masbatenyo	púknit	to remove, detach, take away
10. Aklanon	sábnit	to snatch, seize, grab

*-nut		husk, fiber (dbl. *-but₋₃)
1. PMP	*benut	coconut husk
2. PMP	*bunut	coconut husk
3. PMP	*Runut	plant fibers
4. BM	banut	fibrous husk of coconut
5. Pangasinan	lánot	thick fiber
6. Aklanon	paknot	bark of a tree (used to make fibers)
7. Nggela	penu	coconut outer husk
8. Itbayaten	sanot	strong fiber
9. Sa'a	unu	fibrous spathe of a coconut frond, used for straining
10. Isneg	xónut	the hair that protects the young shoots of fan palms

Note: Also Keley-i *labənət* 'fibrous structures in plants, particularly banana plants'.

The ñ-roots

*-ñam		savory, tasty
1. PWMP	*kiñam	to try, taste
2. PAN	*ñamñam	tasty, delicious
3. PAN	*tañam	to try, taste
4. Kankanaey	ínam	palatable, savory
5. Javanese	kəñam	a taste; a sample
6. Melanau (Matu)	kuñam	to taste
7. Melanau (Mukah)	ñam	taste
8. Bikol	sunám	cloying, satiating
9. Ida'an Begak	tinam	to try

Note: Also Western Bukidnon Manobo *naʔam-naʔam* 'to taste food to see if it is good'.

*-ñat		to stretch (dbl. *-ñet, *-ñit, *-ñut)
1. PWMP	*bañat	to stretch
2. PPH	*beh(e)ñat	to stretch
3. PWMP	*heñat	to stretch
4. PPH	*hik(e)ñat	to stretch
5. PPH	*huñat	to stretch, straighten out
6. PWMP	*íñat	to stretch, pull s.t. that stretches
7. PWMP	*keñat	tough, elastic, of flesh
8. PPH	*kuñat	tough, rubbery, elastic
9. PPH	*uñat	to stretch, straighten out
10. Kayan	añat	to stretch

(continued)

11. Bontok	bínat	to stretch, as an elastic band or a sweater; pull tight, as a loose thread
12. Dusun (Kadazan)	hanat	to stretch
13. Abaknon	hinat	to stretch
14. WBM	kanat	to stretch
15. Ibatan	lonat	to straighten, stretch out s.t.
16. BM	nanat	to stretch (a rope, piece of material, elastic)
17. Pangasinan	pínat	to stretch
18. Cebuano	úgnat	to stretch out s.t. somewhat elastic

*-ñaw.$_1$		to melt, liquefy
1. PMP	*lañaw	to melt, liquefy
2. PWMP	*luñaw	mud, muddy
3. PPH	*ñawñáw	to dissolve s.t. in water
4. PWMP	*Ruñaw	to melt, liquefy
5. PPH	*túñaw	to melt, dissolve
6. Binukid	bunaw	to add liquid to a staple food to make it soft
7. Paiwan	djerenaw	to dissolve, melt
8. Maranao	mbolanaw	to melt, disintegrate; die
9. Maranao	sənaw	to melt, disintegrate

*-ñaw.$_2$		to wash, bathe, rinse
1. PPH	*bal(e)ñaw	to rinse, rinse off
2. PAN	*bañaw	to wash the body
3. PS	*deno	to bathe
4. PWMP	*huñaw	to wash, wash off
5. PAN	*ñawñaw	to rinse, wash
6. PAN	*Señaw	to wash
7. PAN	*Siñaw	to wash
8. Pangasinan	bəgnáw	to rinse
9. Maranao	boanaw	to wash
10. Ilokano	bugnáw	rinse
11. Bikol	bulnáw	to rinse off (as soap, dirt)
12. Bunun	danav	to wash (face, oneself)
13. Dumagat (Casiguran)	gúnaw	to put in water and swish around
14. Kapampangan	kañaw	to rinse; rinsing, washing (Bergaño)
15. Saisiyat	kapranaw	towel for bathing
16. Dumagat (Casiguran)	lábnaw	to rinse (as scraps of food from plate, or soap from cloth)

(continued)

17. Ngadha	leno	to enter the water, bathe the body
18. Aklanon	libánaw	to wash one's hands/feet
19. Kavalan	rinaw	to wash tableware (bowls, cups, etc.)
20. Ilokano	tabnáw	to plunge, cast oneself into water
21. Keley-i	təlipnaw	to rinse dishes, pots or pans
22. Puyuma (Tamalakaw)	tilaw	to rinse

*-ñeb		to dive; sink, disappear under water (dbl. *-nab, *-ñej, *-ñep)
1. PWMP	*ce(R)ñeb	to dive, submerge
2. PAN	*leñeb	to disappear under water
3. PPH	*reñeb	to sink, soak into
4. PAN	*teñeb	to submerge
5. Maranao	galənəb	to sink
6. Ilokano	nebneb	to sink, founder (of ships)
7. Maranao	panəb	to fish for shellfish, dive
8. Ilokano	ráneb	to submerge in water
9. Maranao	salənəb	to dip, sink
10. Maranao	sənəb	to dive, immerse a person in water
11. Maranao	sinəb	to dive
12. Itbayaten	so?nəb	to dive into water
13. Ilokano	táneb	to submerge partly

Note: Also Amis lənəg 'to sink'.

*-ñej		to submerge, drown (dbl. *-nab, *-ñeb, *-ñep)
1. PMP	*keñej	to sink or drown
2. PWMP	*leñej	to sink, disappear under water
3. PMP	*teñej	to sink; to set, of the sun
4. Yami	aned	to sink
5. Lun Dayeh	ened	a drowning
6. Maranao	galənəd	to sink
7. Itbayaten	humnəd	to sink
8. Hawu	keña	to dive, plunge
9. Kanowit	luñet	to sink, drown
10. Kenyah	meñet	to drown
11. DPB	nəŋnəŋ	thrown into the water
12. Bikol	sulnód	to set (of sun, moon, stars); to disappear from view
13. Cebuano	túsnud	to dip, immerse a part of s.t.

Note: Also PPH *luñuj 'to sink'.

*-ñep		to dive; sink, disappear under water (dbl. *-nab, *-ñeb, *-ñej)
1. PMP	*eñep	sunken, submerged
2. PWMP	*leñep	to disappear under water
3. Isneg	dannáp	a flood
4. Lun Dayeh	lanep	bobbing up and down (as when a person drowns); disappear (under water)
5. Kayan	lemñep	to be swept away by a flood
6. Kankanaey	nənəp	to submerge; to be immersed
7. Bikol	sánop	flooded, inundated
8. Bikol	sulnóp	to set (of sun, moon, stars); to disappear from view
9. Kelabit	tanep	to sink, set, disappear from view
10. Puyuma	tənəp	to submerge, send to the bottom
11. Itbayaten	tonəp	idea of being almost fully immersed in a body of water

Note: Also Bikol *sínap* 'referring to the gradual covering of the land by the rising tide', Ngaju Dayak *harenep* 'to sink, descend'.

*-ñet		stretchy, elastic (dbl. *-ñat, *-ñit, *-ñut)
1. Dumagat (Casiguran)	kəlnét	tough (of meat)
2. Ibatan	lānet	to stretch out its body full length after being coiled or in crouching position (of a snake, person)
3. Ibatan	nətnət	to stretch out rope, etc.
4. Ilokano	pennét	tough meat; elastic tissue of quadrupeds
5. Itbayaten	vorinəʔnət	stretching apart, to stretch out (as rubber)
6. Ibatan	yənənət	to stretch oneself

*-ñit		stretchy, elastic (dbl. *-ñat, *-ñet, *-ñut)
1. Kelabit	binit	stretchable, stretched
2. Aklanon	ganít	tough, hard to chew
3. Cebuano	húnit	tough and resilient (as pork rinds)
4. Bikol	ʔignít	to stretch something
5. Tboli	lenét	to stretch, as clothes

*-ñut		stretchy; elastic; springy (dbl. *-ñat, *-ñet, *-ñit)
1. PWMP	*lañut	tough, as meat
2. PWMP	*uñut	to stretch oneself
3. Ilokano	ánut	meat full of sinews
4. Iban	bəñut	shake, quake, quiver, be springy, with give in it (as a weak beam)
5. Cebuano	kínut	for food to be tough and hard to break, but flexible
6. Hanunóo	kúnut	flexibility, pliability
7. Kankanaey	kuynút	flexible, pliable, pliant
8. Tboli	not ~ nut	the elasticity of s.t.
9. Hiligaynon	ʔuganút	tough (meat)

Note: Also Mansaka ganod 'to stretch; to expand'.

The *N-roots

*-NaR		ray of light
1. PPH	*banaʔaR	rays of rising sun
2. PWMP	*benaR	sunlight
3. PPH	*inaR	rays of the sun
4. PAN	*saheNaR	to shine, of the sun
5. PAN	*siNaR	ray of light
6. PAN	*suNaR	light, radiance
7. DPB	binar	light, radiance
8. Soboyo	dina	sun
9. Maranao	inanag	radiant heat
10. Kamarian	kina	glitter, sparkle, shine
11. Bikol	liwánag	light, daylight
12. Maranao	matamenag	white, clear, bright
13. Manggarai	ninar	somewhat illuminated (of the night)
14. Melanau (Mukah)	panah	ray of light
15. Paiwan	pa-qulʸa	to make light
16. Ilokano	silnág	shine, sparkle, flash, glisten, gleam
17. Chamorro	somnak	sunshine, sunlight, ray (from the sun)
18. Maranao	tamnag	light
19. Paiwan	telʸar	brightness; light (natural)
20. BM	ulinag	shine, radiance (as of sunlight)

*-Naw		clear, pure (of water)
1. PAN	*CiNaw	clear, pure, of water
2. PMP	*linaw	calm, still, of the surface of water
3. PPH	*tenaw	clear, of liquid free of impurities
4. Bikol	galínaw	to see s.t. beneath the surface of the water
5. Ilokano	litnáw	clearness, purity, transparency

*-Neb		door, doorway
1. PAN	*qaNeb	doorway, door
2. PAN	*qeNeb	door(way); close a door
3. PAN	*qiNeb	door; to close
4. Dumagat (Casiguran)	gánəb	door; to close a door
5. Sa'a	hono	to shut, to bar; a door
6. Ayta Abellen	lənəb	to close; door
7. Yami	panəb	door (Ferrell 1969)
8. Balinese	unəb	carved wooden panel put over a doorway as an ornament; to lock

*-Neŋ		to stare, look fixedly (dbl. *Nuŋ, *-teŋ₋₂)
1. PAN	*Neŋ	to look, see
2. PAN	*NeŋNeŋ	to stare, look fixedly
3. PWMP	*sineŋ	to look at closely, stare intently
4. Kayan	belaneŋ	to stare at; open, opened, of eyes
5. Kayan	ineŋ	to see, look at
6. Kayan	jebaneŋ	to open the eyes wide
7. Kenyah	naneŋ	to stare at
8. Kenyah	nekeneŋ	looking on idly
9. Kenyah	ŋeneŋ	to look at, investigate
10. Murut (Timugon)	mo-ponoŋ	be viewed, stared at
11. Uma	-rono	to look at
12. Kayan	sebaneŋ	to open one's eyes
13. Melanau (Mukah)	seneŋ	to stare
14. Tboli	teneŋ	to look at s.t. with awe, wonder, without blinking
15. Fijian	wānono	to stare at

*-NiC		to skin, peel off
1. PAN	*baNiC	to skin, flay
2. PS	*benit	to peel off (rind, bark)
3. PMP	*panit	to skin, flay
4. PAN	*qaNiC	skin, hide
5. Rembong	bunit	to peel the stalks of vegetables
6. Iban	kanit	to strip off with the teeth
7. Ida'an Begak	lanit	to peel off skin
8. Bikol	ʔúnit	edible skin, as that of a roasted pig

*-NuC		to pull out, uproot (dbl. *-buC, *-guC, *-zut)
1. PAN	*beNuC	to pull out, extract
2. PPH	*búnut	pulling out, uprooting
3. PPH	*gánut	to pull out, as grass or a tooth
4. PPH	*rab(e)nút	to grab and pull
5. PPH	*sabunut	to pull by the hair; to pull out hair
6. Aklanon	bugnót	to yank, pull out
7. Waray-Waray	bulnót	the act of pulling up, pulling out
8. Ibaloy	daʔnot	to pull up a plant by the roots
9. Amis	faʔnot	to molt
10. Amis	foʔnot	to uproot
11. Sambal (Botolan)	habónot	to pull the hair of a person one is angry with
12. Itbayaten	kohnot	pulling out (as nails, saber); leaving
13. Aklanon	kúsnot	to playfully grab at another, snatch at
14. Kavalan	m-runut	to pull out hair
15. Keley-i	uknut	to pull out (as a bolo from its scabbard)
16. Itbayaten	vornot	pulling out (esp. weeds)

*-Nuŋ		to stare, look fixedly (dbl. *Neŋ, *-teŋ$_{-2}$)
1. PWMP	*nuŋnuŋ	to stare, look fixedly
2. Iban	m-əncənoŋ	open and staring (of eyes)
3. Iban	gə-gənoŋ	to stare
4. Kankanaey	ónoŋ	to have an eye upon; to look eagerly at; to watch covetously
5. Malay	rənoŋ	to look fixedly at anything

The *ŋ-roots

*-ŋa		to gape, open the mouth wide (dbl. *-ŋab, *-ŋaŋ, *-ŋap, *-ŋaq)
1. PMP	*aŋa	wide open (of the mouth)
2. PWMP	*gaŋa	to open the mouth
3. PMP	*kaŋa	to be open, as the mouth
4. PAN	*ŋaŋa	to open the mouth
5. PAN	*paŋa	fork of a branch, bifurcation
6. POC	*paŋa	to gape, be open
7. PMP	*peŋa	fork of a branch
8. PMP	*Raŋa	to open the mouth wide
9. PMP	*saŋa	bifurcation, fork of a branch
10. PAN	*taŋa	to open wide
11. Lau	afaŋa	open wide, gape; uncover
12. Malay	bəlaŋa	wide-mouthed cooking pot
13. Malay	bəŋa	amazed; dumbfounded
14. Gilbertese	beŋa-beŋa	large opening, separation; separated, wide apart
15. Cebuano	buŋáŋa	for s.t. to be open wide or ajar (as the mouth falling open in surprise)
16. Uma	daŋa	handspan
17. Aklanon	dáŋadaŋá	to pant, breathe quickly
18. Javanese	əŋa	opening
19. Rotinese	fiŋa	to force the mouth open, as with a sick person to give him medicine
20. Iban	jaŋaʔ	angle formed by fork in a stick, junction of streams or roads, etc.
21. Karo Batak	jəraŋa	to stand apart, as flowers in a bunch
22. Manggarai	keŋa	to seize with open mouth (of an animal); eat with open mouth
23. Kenyah (Long Atun)	pəbəŋa	wide open, of the mouth
25. Kankanaey	sakgaŋá	branched horns (of deer)
26. Malagasy	sóŋga	having the upper lip turned upwards
27. Keley-i	təŋa	to amaze, or to be amazed or surprised
28. Ngadha	viŋa vaŋa	to open the mouth wide

Note: Also Cebuano ŋáŋha 'be open-mouthed with sudden surprise'. Final glottal stop in Iban is assumed to be historically secondary (cf. Blust 2013:574–81).

*-ŋab		to gape; open, of the mouth (dbl. *-ŋa, *-ŋaŋ, *-ŋap, *-ŋaq)
1. PMP	*qaŋab	to gape, open the mouth wide
2. Tiruray	kaŋab	to take a large bite out of s.t.
3. Sangir	laŋabeʔ	gaping; wailing

(continued)

4. Cebuano	ŋábŋab	for a wound or an opening to become big or wide
5. Maranao	oŋab	mouth that is naturally open
6. Itbayaten	saŋab	greed, voraciousness, gluttony
7. Sangir	taŋabeʔ	wide open (of a large pair of scissors, a wide hole, a gaping wound, etc.)
8. Cebuano	tiŋábŋab	to be gaping with a huge opening or hole

*-ŋaC	anger, irritation (dbl. *-get, *-git, *-gut, *-ŋeC, *-ŋiC, *-ŋuC)	
1. Balinese	jəŋat	to put on a sour face; speak sternly, bitingly
2. Lun Dayeh	raŋat	acting in an angry manner
3. Paiwan	seŋats	one who characteristically dislikes things
4. Kayan (Busang)	siŋat	to speak in irritation

*-ŋag	to look upward	
1. Maranao	aŋag	to look up to imploringly
2. Malay	caŋak	a sudden startled upward look
3. Sasak	cəŋak	to look up (from one's work)
4. Malay	coŋak	tilted up (of the face); chin up
5. Malay	daŋak	to hold the head back
6. Malay	doŋak	tilting the head forward and up, of a man looking skyward
7. Malay	jəlaŋak	to throw back the head, e.g. for watching an airplane
8. Tiruray	ləŋag	to look up
9. Maranao	litaŋag	to look up
10. Maranao	tiŋag	to look up or to look up to

Note: The final consonant in Javanese dəŋak, ḍəŋak 'facing upward, craning the neck', laŋak '(puppet with an) upturned gaze', and iŋakuŋak 'crane the neck in an effort to see' can only reflect *k. Since Malay and Sasak -k reflect both *-g and *-k, it is possible that the forms cited from these languages contain a root *-ŋak, and are not to be compared directly with the Philippine material.

*-ŋak		to screech, howl (dbl. *-ŋek, *-ŋik, *-ŋuk)
1. PMP	*heŋak	out of breath
2. PMP	*ŋak	raucous sound
3. PMP	*ŋakŋak	raucous sound
4. Javanese	bəŋak-bəŋak	to shout repeatedly
5. Old Javanese	burəŋak	to scream
6. Old Javanese	caŋak	heron sp. (onom.)
7. Old Javanese	cəŋak	the cry of a *caŋak* bird (heron sp.)
8. Kankanaey	galakúŋak	to be talking, jabbering, babbling, chattering (of many people together)
9. Cebuano	híŋak	to breathe deeply and rapidly
10. Banjarese	huŋak	to gasp for breath
11. Manggarai	kaŋak	to howl, whine (of a dog that is struck)
12. Manggarai	keŋak	(of a baby) cry
13. Toba Batak	laŋak	to honk (geese)
14. Kankanaey	ŋáŋak	to cry, weep (used only in tales)
15. Hanunóo	paŋák	barking of dogs
16. Bikol	paragŋák	a loud, menacing cry or shout
17. Iban	pərəŋak	to shout at in anger
18. Bikol	pusŋák	to guffaw; to laugh heartily
19. Malay	səŋak-səŋok	to sob, sobbing speech
20. Bikol	suŋák-suŋák	the sound of gasping or panting
21. Bikol	ʔusŋák	to snort

*-ŋaŋ		amazed; gaping (dbl. *-ŋa, *-ŋab, *-ŋap, *-ŋaq)
1. PMP	*beŋaŋ	to gape, open the mouth wide; be dazzled, amazed
2. PMP	*ciŋaŋ	to gape, open the mouth
3. PWMP	*leŋaŋ	amazement, astonishment
4. PWMP	*seŋaŋ	startled
5. PWMP	*zeŋaŋ	open-mouthed
6. Old Javanese	aŋaŋ	wide open, gaping
7. Sasak	biŋaŋ	harelip
8. Sundanese	buŋaŋaŋ	wide open, wide opening
9. Old Javanese	burəŋaŋ	amazed, aghast
10. Malay	cəŋaŋ	bewilderment, amazement, astonishment
11. Rembong	laŋaŋ	surprised, startled
12. Buli	paŋaŋ	to gape, yawn
13. Iban	raŋaŋ-raŋaŋ	with claws open (of scorpions or crabs)
14. Ilokano	suŋáŋ	to have the upper jaw and teeth protruding
15. Manggarai	teŋaŋ	dizzy; confused
16. Ilokano	tuŋŋáŋ	fool, simpleton, idiot, imbecile

*-ŋap		to open, of the mouth; gaping (dbl. *-ŋa, *-ŋab, *-ŋaŋ, *-ŋaq)
1. PMP	*beŋap	surprised, amazed
2. PMP	*ceŋap	to seize in the mouth or beak
3. PWMP	*ciŋap	to catch one's breath
4. PMP	*eŋap	to gasp for breath
5. PMP	*qaŋap	to gape, open the mouth wide
6. PMP	*uŋap	to open the mouth wide
7. Sundanese	calaŋap	to gape
8. Javanese	caŋap	a wide opening
9. Javanese	craŋap	(of mouths) wide open
10. Malay	cuŋap-caŋip	puffing and blowing; panting
11. Pangasinan	doŋáp	to breathe heavily from illness or exhaustion
12. Ayta Abellen	hiŋap	to be short of breath
13. Sambal (Botolan)	hoŋáŋap	to gasp for breath
14. Iban	jəŋap	gasping, fighting for breath, in a dying state
15. Pangasinan	laŋáp	to force a dog to eat by holding its mouth open and inserting food
16. Iban	lələŋap	gaping (?); open-mouthed (?)
17. Iban	məŋap	to be surprised, stare (in amazement)
18. Minangkabau	ŋaŋap	to snap at flies, of a dog
19. Ngaju Dayak	raŋa-raŋap	to open (mouths of crocodiles, etc.)

*-ŋaq		to gape, open the mouth wide to open, of the mouth; gaping (dbl. *-ŋa, *-ŋab, *-ŋaŋ, *-ŋap)
1. PMP	*aŋaq	to open the mouth wide
2. PWMP	*balaŋa(q)	inattentive
3. PMP	*baŋaq	open mouth
4. PMP	*beŋaq	to open, as the mouth
5. PPH	*buŋáŋaq	to open the mouth wide
6. PWMP	*daŋaq	wide open, of the mouth
7. PWMP	*eŋaq	wide open (of the mouth)
8. PMP	*laŋaq	to gape, wide open
9. PWMP	*ŋaqŋaq	to open the mouth, gape
10. PAN	*paŋaq	forked, pronged
11. PWMP	*taŋaŋaq	to open the mouth wide; gape
12. PWMP	*tiŋaqŋaq	to open wide
13. DPB	dalŋah	wide open, as the mouth
14. Aklanon	iŋaʔ	to bellow (carabao)
15. Hawu	kejaŋa	branches of a tree
16. Bikol	liŋáŋaʔ	wide-mouthed (as a pitcher or jar)
17. Melanau (Mukah)	maŋaʔ	crack, fissure
18. Sundanese	raŋah	bit (for horse's mouth)
19. Maranao	siŋaʔ	to laugh

(continued)

20. Karo Batak	talŋah	to open very wide (door)
21. Cebuano	tásŋaʔ	to grin idiotically
22. Tagalog	tuŋáŋaʔ	open-mouthed, agape

*-ŋar		to howl, shout, scream
1. PMP	*ŋarŋar	to talk loudly or angrily; howl
2. Bontok	ʔáŋal	to answer back angrily; to challenge
3. Balinese	saŋar	to yell, scream
4. Yamdena	taŋar	to bark, of dogs
5. Hiligaynon	uŋál	to drawl, howl

Note: Also PMP *iŋaR 'loud, unpleasant noise', PMP *quŋal 'mournful howl of a dog'.

*-ŋaw$_{-1}$		confused, disoriented, lost (dbl. *-gaw, *-taw$_{-2}$)
1. Ifugaw (Batad)	amuŋaw	for s.o. to continuously act foolishly, be confused about s.t.
2. Binukid	buŋaw	senile, absentminded, forgetful, confused (due to old age)
3. Kavalan	mRimseŋŋaw	to confuse, bewilder
4. Bikol	muŋáw-muŋáw	still sleepy after waking up; dazed, groggy, stunned
5. Ibaloy	ŋalawŋaw	confused, disorderly, noisy—esp. of a crowd with some shouting, laughing, fighting (as the revelry of drunks)
6. Bontok	ŋawŋaw	to be mixed up mentally, to be confused
7. Bikol	ruŋáw	crazy, insane, mad; a fool, madman
8. Ibaloy	saŋaw	to be off course, lost, confused (as one lost in the forest)
9. Ifugaw (Batad)	teŋteŋaw	for a place spirit to cause a person, animal, thing to become lost, e.g. in a forest

*-ŋaw$_{-2}$		fly (generic for housefly and others)
1. Melanau (Mukah)	bəlaŋaw	housefly
2. Ilokano	buŋáw	kind of large green dragonfly
3. Sangir	həmbaŋo	firefly
4. Tagalog	láŋaw	housefly
5. Iban	pəluŋaw	a poisonous blood-sucking fly
6. Puyuma (Tamalakaw)	veRaRŋaw	bluebottle fly

*-ŋaw₋₃		leaking air, vapor
1. PWMP	*seŋaw	vapor, steam
2. PPH	*siŋaw	vapor, leaking air, fumes
3. PPH	*suŋáw	to evaporate, as vapor rising from the ground after rain
4. Bikol	ʔalisŋáw	to give off, emit, issue; to escape (vapors, smells); to evaporate
5. Bikol	haŋáw	breath

*-ŋaw₋₄		rice bug, insect destructive to crops
1. PAN	*baŋaw	paddy bug, foul-smelling insect that preys on rice
2. PWMP	*tuŋaw	tiny red itch mite
3. PWMP	*zaŋaw	stinkbug destructive to rice
4. Malay	aŋaw	small jungle tick
5. Malay	cənaŋaw	foul-smelling bug that preys on rice
6. Malay	paŋaw	foul-smelling bug that preys on rice
7. Hanunóo	púŋaw	taro beetle
8. Tiruray	taŋəw	a rice bug: *Leptocorisa acuta* (Thunberg)
9. Rejang	tənaŋəw	paddy stalk borer

*-ŋeC		angry; to gnash the teeth (dbl. *-get, *-git, *-gut, *-ŋaC, *-ŋiC, *-ŋuC)
1. PWMP	*buŋet	anger, angry, irritated, annoyed
2. PPH	*haŋet	to gnaw, chew on
3. PAN	*ŋeCŋeC	to gnaw, nibble
4. PMP	*qaŋet	anger
5. PMP	*qiŋet	angry, upset
6. PAN	*reŋeC	angry, annoyed
7. PAN	*ReŋeC	angry, annoyed
8. PPH	*suŋet	irritated, angry
9. PWMP	*uŋet	anger
10. Itbayaten	ahŋət	gnawing, biting
11. Ida'an Begak	bəriŋot	be angry without showing it
12. Tombonuwo	o-biŋuŋot	always angry
13. Aklanon	dalipuŋót	frustrated; out of sorts, crabby
14. Itneg (Binongan)	iŋit	anger
15. Maranao	gaŋət	to crush with teeth, chew
16. Mansaka	iŋutiŋut	to be angry with s.o.
17. Ifugaw	kayŋót	to bite on one's lower lip thereby to show one's anger or displeasure

(continued)

18. Kankanaey	kiŋə́t	choleric, irascible
19. Manobo (Ilianen)	laŋit	anger
20. Kankanaey	liŋə́ʔət	to grate, produce a harsh sound
21. Yami	niŋet	frown
22. Dumagat (Casiguran)	ŋahitiŋət	the cracking sound of a person's teeth as he grinds them together
23. Kankanaey	ŋaŋŋə́t	to chide, scold, speak harshly to s.o.
24. Itbayaten	ŋarətŋət	grinding teeth when asleep
25. Kankanaey	ŋət	to creak, gnash (doors, teeth, etc.)
26. Bikol	raʔŋót	to gnaw on
27. Iban	riŋat	angry, annoyed
28. Isneg	ruŋát	anger
29. Ilokano	rupaŋət	to frown, scowl, look angry
30. Itbayaten	saŋət	making faces when one dislikes taste, frowning
31. Ilokano	siduŋét	a frown
32. Dusun (Kadazan)	soŋot	angry
33. Agutaynen	toriŋet	to scowl with the eyebrows bunched up; to scowl at a person
34. Kayan	tuŋet	scolding, being angry
35. Paiwan	vereŋets	wrinkled; scowling (face)
36. Aklanon	yáŋot	to hate, despise

Note: Also PAN *ŋitŋit 'to gnaw', PMP *ŋutŋut 'to gnaw'.

*-ŋek		to grunt, groan (dbl. *-ŋak, *-ŋik, *-ŋuk)
1. PMP	*ŋek	a grunt
2. PMP	*ŋekŋek	to mumble, etc.
3. PWMP	*seŋek	difficulty in breathing
4. Kankanaey	ʔataŋŋə́k	to resound loudly, ring
5. Kankanaey	ŋaʔəkŋə́k	to bubble, as boiling rice

*-ŋel		deaf
1. PMP	*beŋel	deaf
2. PWMP	*biŋel	acting deaf, refusing to listen
3. PWMP	*iŋel	hard of hearing
4. Palawano	baŋel	deafened by
5. Kenyah	deŋen	deaf
6. Soboyo	poŋo	deaf
7. Madurese	teŋel	deaf
8. Agutaynen	tiŋel	deaf, hard of hearing
9. Sasak	toŋəl	deaf

*-ŋeŋ		to buzz, hum (dbl. *-ŋiŋ, *-ŋuŋ)
1. PWMP	*eŋeŋ	to buzz, hum
2. PMP	*ŋeŋ	a buzzing, humming
3. PMP	*ŋe(ŋ)ŋeŋ	to buzz, hum
4. Itbayaten	arəŋəŋ	murmuring
5. Malay	bəŋaŋ	drumming or buzzing in the ear
6. Javanese	bəŋəŋəŋ	to buzz, whine, hum
7. Toba Batak	loŋoŋ	buzz, drone, of mosquitoes
8. Javanese	rəŋəŋ	soft singing or humming voice

*-ŋeR		to hear; noise
1. PWMP	*beŋeR	deafened
2. PAN	*deŋeR	to hear; sound
3. PWMP	*diŋeR	to hear
4. PAN	*kiŋeR	to hear
5. PWMP	*qiŋeR	loud, unpleasant noise
6. PPH	*tiŋeR	sound; voice
7. Malay	baŋar	clamor
8. Kankanaey	batúŋəg	dull of hearing; deaf
9. Sasak	biŋər	deafened by noise
10. Keley-i	kikkiŋŋəl	an echo sound in the forest or bush
11. Mansaka	liŋəg	to lose hearing; to hear poorly
12. Bikol	ŋalimbuŋóg	to turn a deaf ear to; to act like you don't hear s.o.
13. Toba Batak	piŋor	full of sounds, so that it cannot hear clearly (of the ear)

Note: Also Agutaynen *iŋal* 'noise'.

*-ŋet		sweat, perspiration
1. PWMP	*liŋet	sweat, perspiration
2. Tboli	iŋet	perspiration
3. Bikol	gaʔnót	sweat, perspiration
4. Malay	kəriŋat	sweat, perspiration

Note: Also Yakan *soŋot* 'sweat, perspiration'.

*-ŋiC		anger, irritation (dbl. *-get, *-git, *-gut, *-ŋaC, *-ŋeC, *-ŋuC)
1. PWMP	*biŋit	moody, irritable
2. PMP	*haŋit	anger

(continued)

3. PAN	*ŋiCŋiC	to show annoyance or irritation
4. PAN	*quSeŋiC	anger; angry
5. PAN	*reŋiC	to grimace at pain
6. PWMP	*saŋit	anger; angry
7. PWMP	*seŋit	irritated, annoyed
8. PWMP	*siŋit	violent emotion
9. PPH	*súŋit	irritable, in a bad mood
10. PAN	*teŋiC	to grimace at pain
11. PWMP	*uŋit	to tease, vex, annoy
12. Cebuano	aliŋít	short-tempered, irritable
13. Kankanaey	báŋit	to tease, plague, trouble, vex, annoy
14. Maranao	bəŋit	quarrelsome, irritable
15. Kankanaey	buláŋit	to tease; to plague; to trouble; to annoy; to pester (used only in tales)
16. Kankanaey	gáŋit	to provoke, challenge, defy
17. Aklanon	iŋít	crabby, out of sorts, in a bad mood
18. Iban	ləŋit	(of words) unkind
19. Kankanaey	miŋít	to jabber, chatter, as when slightly angry
20 Sangir	peŋi?	anger
21. Maranao	raŋit	anger
22. Bikol	riŋít	to feel out of sorts because of one's menstrual period or pregnancy
23. Kavalan	RŋiRŋit	to gnash the teeth
24. Kankanaey	saŋíŋit	to resent, be offended, get angry
25. Itbayaten	saroñit	being cruel or very impatient
26. Paiwan	vireŋits	to grimace at pain
27. Paiwan	viteŋits	to make faces

*-ŋik		a shrill throaty sound (dbl. *-ŋak, *-ŋek, *-ŋuk)
1. PWMP	*eŋik	high-pitched sound of complaint
2. PMP	*ŋik	a squeal, screech, shriek
3. PMP	*ŋikŋik	to squeal, screech, shriek
4. Javanese	bəŋak-bəŋék	to keep shouting or crying out
5. Old Javanese	burəŋik	to shriek
6. Balinese	cəŋik	to make a mournful sound: ŋik
7. Sundanese	dəŋek	letting out a scream or shriek
8. Javanese	éŋék	playing of a stringed instrument
9. Iban	ŋaŋik	to shriek
10. Iban	rəŋik	(of children) crying and pestering; to gasp, grunt, squeal
11. Iban	riŋik	to giggle

(continued)

12. Sundanese	səŋek	a shrill sound; let out a shriek
13. Malay	taŋek	a nasal twang
14. Dusun (Kadazan)	toŋik	high-pitched sound
15. Ilokano	uŋík	the shrill cry of a pig, as when caught by men

*-ŋiŋ		a shrill buzz or hum; to ring (dbl. *-ŋeŋ, *-ŋuŋ)
1. PMP	*ŋiŋ	a buzzing, humming
2. PMP	*ŋi(ŋ)ŋiŋ	to buzz, hum
3. Malay	ceŋeŋ	continuous whining (of infants)
4. Malay	dəŋiŋ	to buzz (of the ear)
5. Malay	doŋeŋ	to tell a story in recitative; the whining of a young child
6. Puyuma	iŋiŋ	the whining of a dog
7. Iban	mərəŋiŋ	to hum, buzz; shrill, like a cicada
8. Lun Dayeh	reŋiŋ	ringing noise in the ears
	fe-reŋiŋ	the sound made by a swarm of mosquitos
9. Dusun (Kadazan)	soŋiŋ	to tinkle
10. Ngadha	viŋi	to ring, tinkle, clink

*-ŋis.₁		to bare the teeth
1. PMP	*beŋis	to grin widely, showing the teeth
2. PMP	*ŋisŋis	to grin, show the teeth
3. PMP	*riŋis	to bare the teeth in smiling or laughing
4. PMP	*waŋis	to bare the teeth
5. PMP	*wiŋis	to bare the teeth
6. Pangasinan	buŋis	harelip (McFarland 1977)
7. Javanese	cəŋiŋis	a jeering grin
8. Manggarai	ciŋis	to grin, showing the teeth
9. Old Javanese	duŋis	with a short (curled) upper lip
10. Old Javanese	iŋis	bared, visible (teeth, etc.), to bare the teeth, grin, smile
11. Manggarai	nciŋis	of the teeth, to be visible because of a curvature of the upper lip
12. Makassarese	paʔŋisiʔ	to grin (as a monkey that shows its teeth)
13. Sangir	taŋisi	to bare the teeth (as children when they laugh)
14. Amis	toŋiŋis	to pull on meat with teeth, tearing it apart to eat
15. Bikol	ʔuŋís	to grimace, snarl

*-ŋis.₂		cruel
1. PWMP	*baŋis	cruel
2. PMP	*beŋis	angry; quarrelsome
3. PWMP	*riŋis	cruel
4. Balinese	jəŋis	to put on a stern expression, be hostile
5. Bikol	ʔuŋís	choleric; to abhor, despise, detest, hate, loathe

Note: Possibly identical to *-ŋis.₁.

*-ŋit		to cry out, shriek
1. PMP	*beŋit	to cry out, as in in pain or fear
2. PPH	*heŋit	to laugh; laughter
3. Dumagat (Casiguran)	kəlaŋet	to yell, to hoot, to scream (an extremely loud yell of excitement)
4. Itbayaten	ñiitñit	squealing sound or cry (of pigs)
5. Rejang	tətəŋit	to laugh, squirm

*-ŋuC		to mumble, mutter; feel annoyed (dbl. *-get, *-git, *-gut, *-ŋaC, *-ŋeC, *-ŋiC)
1. PWMP	*deŋut	restless, impatient, annoyed
2. PMP	*eŋut	to complain loudly
3. PMP	*huŋut	to grumble, sulk
4. PWMP	*meŋut	angry, grumbling
5. PMP	*ŋut	to grunt
6. PWMP	*ŋutŋut	to mumble, whimper
7. PMP	*reŋut	to frown (as in anger)
8. PWMP	*uŋut	to mumble angrily, grumble
9. Keley-i	aŋut	to look cross; to frown, pout or sulk
10. Waray-Waray	aroŋót	disgusted; exasperated; the act of crying, sobbing, or expressing a complaining tone
11. Kapampangan	baŋuŋut	to have a nightmare, be affected by a seizure during sleep
12. Ngadha	beŋu	sullen, sour, angry, to have a grudge; growl, grumble
13. Toba Batak	biruŋut	morose, sulky in appearance
14. Aklanon	buriŋót	bad-humored, moody, out-of-sorts
15. Pangasinan	iŋóŋot	annoyance
16. Mansaka	iŋutiŋut	to be angry with s.o.
17. Itbayaten	laŋot	pouting

(continued)

18. Iban	ləŋə-ləŋut	looking angry
19. Kankanaey	ŋalutúŋut	to murmur, grumble, mutter
20. DPB	riŋut	frowning; sour face
21. Iban	suŋut	to murmur, grumble, bear a grudge against
22. Kayan	teŋut	the sound of a distant roar or hum
23. Paiwan	vuteŋuts	to make a wry face

*-ŋuk		a deep throaty sound (dbl. *-ŋak, *-ŋek, *-ŋik)
1. PMP	*ŋuk	a grunt, moan, etc.
2. PMP	*ŋukŋuk	to grunt, moan, etc.
3. Javanese	bəŋok-bəŋok	to shout repeatedly
4. Old Javanese	cəŋuk	of the cry (songs? warble?) of a quail
5. Aklanon	iŋók	to grunt (like a pig)
6. Ngaju Dayak	nanteŋok	to drink water
7. Kankanaey	ŋóŋok	to grunt
8. Maranao	rəŋok	to snore
9. Malay	səŋok	to sob

*-ŋuŋ		deep buzzing or humming (dbl. *-ŋeŋ, *-ŋiŋ)
1. PMP	*beŋuŋ	a humming sound
2. PMP	*ŋuŋ	a buzzing, humming
3. PMP	*ŋu(ŋ)ŋuŋ	to buzz, to hum
4. Malay	dəŋoŋ	a boom, deep hum
5. Hawu	keŋu	to cry, howl (as a dog)
6. Kayan	meteŋuŋ	the sound of a swarm of bees or other flying insects
7. Iban	rəŋoŋ	to groan, growl; humming, drone, whine

*-ŋur		a low-pitched sound (dbl. *-duR, *-gur, *-kur)
1. PPH	*úŋur	to moan, growl
2. Buli	asŋor	to grunt, of a pig
3. Pangasinan	daŋól	to bark
4. Kelabit	dəŋur	low-pitched, of sound
5. Singhi	ŋur	to growl
6. Pazeh	ŋuruŋur	to scold, to grumble and complain to oneself while walking around
7. Kelabit	piŋur	echo

Note: Also Hanunóo *áŋir* 'the growl of a dog', PAN *ŋerŋer 'growl'.

*-ŋus		snout	
1. PPH	*duŋus	snout, nose	
2. PPH	*eŋus	to sniffle, pant	
3. PMic	*faŋus-	to blow the nose	
4. PWMP	*heŋus	to breathe	
5. PWMP	*iŋus$_1$	to sniffle; snot	
6. PMP	*piŋus	nasal mucus, snot	
7. PMP	*qiŋus	nasal mucus, snot	
8. PPH	*táŋus	pointed or prominent, of the nose	
9. PWMP	*uŋus	snout, beak	
10. Simalur	aŋos	snout, beak	
11. Pazeh	babaŋut	area between the mouth and nose	
12. Tagalog	balúŋos	snout of a fish	
13. Sundanese	baŋus	snout	
14. Javanese	bəŋus	to snort	
15. Arosi	heŋu	to hold the nose	
16. Sundanese	jəŋos	to snort, as a water buffalo that wants to get loose	
17. Tagalog	ŋusŋós	tip of nose	
18. BM	totoŋut	nose	
19. Dusun (Kadazan)	tuoŋus	pig's snout, elephant's trunk	
20. Dusun (Kadazan)	umpoŋus	to snort	

*-ŋuy		to swim
1. PMP	*laŋuy	to swim
2. PAN	*Naŋuy	to swim
3. PWMP	*taŋuy	to swim
4. Tunjung	karaŋoy	to swim
5. Berawan (Long Terawan)	pəlaŋoy	to swim
6. Bintulu	pəriŋuy	to swim

The *p-roots

*-paD		flat, level
1. PMP	*da(m)paD	flat, level
2. PWMP	*de(m)paD	flat and wide
3. PMP	*lapaD	broad and flat
4. PWMP	*sapaD	flat, of things that are expected to be more rounded
5. Malay	papar	flat, smooth, blunt; back of blade (in contrast to edge)

Note: Also PWMP *parpar 'flat, even, level'.

*-pag		to strike, beat (dbl. *-pak₋₃)
1. PWMP	*lempag	to strike, hit
2. PWMP	*pagpag	to slap something hard
3. PWMP	*repag	to slap
4. PPH	*upag	to pound on
5. Kelabit	əpag	the act of slapping
6. Ifugaw	hípag	blow in general, but not with a weapon (e.g. a stick); mostly used of blows with one's fist
7. Kapampangan	kalampág	tapping noise
8. Bikol	laspág	to beat up
9. Mansaka	lompag	to break up (as stone); to tear down (as house)
10. Aklanon	páepag	to hammer using the broad side of anything, like a rock, piece of metal or wood
11. Kankanaey	pagípag	to bang, knock, bounce
12. Ilokano	pampág	to beat, strike repeatedly (rugs, etc.), either with the hand or an instrument
13. Ibaloy	tipag	to strike s.t. against s.t. else
14. Ifugaw	tolpág	to hit s.o. or s.t. accidentally
15. Itbayaten	xopag	striking (with a stick, bat, etc.)

Note: Also Balinese *sapag* 'chop violently at, hew'.

*-pak₋₁		to break, crack, split (dbl. *-bak₋₃)
1. PWMP	*cepak	to crack, split, break
2. PWMP	*epak	to split, break off
3. PWMP	*hepak	to split, break off
4. PWMP	*hupak	to split, peel off
5. PWMP	*ipak	to split
6. PWMP	*kupak	to break open, as a ripe fruit
7. PMP	*lepak	to break, crack off
8. PWMP	*lu(m)pak	to break, split off
9. PMP	*papak	to break, chip off
10. PWMP	*rapak	to break off branches
11. PWMP	*ripak	to split, crack
12. PMP	*Repak	to break to pieces
13. PWMP	*Ri(m)pak	to split, crack
14. PMP	*sapak	to crack, split, break; sound of cracking
15. PWMP	*sepak	to break, snap off at a joint
16. PWMP	*sipak	to crack apart at the joints
17. PMP	*tempak	to split or break off
18. Lun Dayeh	afak	a split; to split, as firewood
19. Karo Batak	gupak	large cleaver with a blunt end
20. Chamorro	hamafak	fragile, brittle, easily broken

(continued)

21. Dusun (Kadazan)	hapak	to crack, split
22. Dusun (Kadazan)	himpak	to split (intr.)
23. Cebuano	ipák	to split, break a piece of something
24. Ibaloy	Kepak	to split along a grain (as coconut husks and bamboo tubes break)
25. Murut (Timugon)	kuripak	to make a cracking sound, as of wood being split hard
26. Tagalog	lapák	torn off; broken off (as tree branches)
27. Kenyah (Long Atun)	pəpak	to split in two
28. Bontok	ʔapák	split or divided, as a forked stick or a tree with a divided trunk
29. Kapampangan	-supák	cracked
30. Maranao	tipak	to break or chip off
31. Ifugaw	tumpák	breaking of anything that can be broken and can no longer be used
32. Toba Batak	ulpak	hit, hew, chop

*-pak.$_2$		to peel bark from a tree (dbl. *-bak)
1. PMP	*pakpak	to peel off bark
2. PWMP	*upak	bark, rind; to peel off bark
3. Banggai	ma-lapak	skinned, peeled off
4. Uma	mo-lepaʔ	skinned, peeled off
5. Balinese	kəlupak	to strip off the leaf-like skin of the bamboo stem or the areca palm
6. Kayan	pak	the fallen sheath from palm frond or bamboo; the husk of sugarcane
	me-pak	to remove the outer sheath of palm frond, or the husk of sugarcane

*-pak.$_3$		to slap, clap (dbl. *bak.$_2$, *-pag, *-pek, *-pik, *puk.$_1$)
1. PWMP	*ca(m)pak	to smack
2. PWMP	*dag(e)pak	loud slap
3. PPH	*dapák	to slap
4. PWMP	*depak	sound of smacking, etc.
5. PMP	*du(m)pak	to strike, collide with
6. PPH	*hampák	to slap, smack
7. PWMP	*ipak	sound of smacking or slapping
8. PMP	*kapak	to flap the wings; sound of flapping
9. PWMP	*kipak	to flap the wings
10. PWMP	*lagepak	to slap; sound of a collision

(continued)

11.	PWMP	*lapak	to smack, slap, slam
12.	PWMP	*lepak	to thud
13.	PWMP	*lipak	to hit, slap
14.	PMP	*pak	the sound of a clap or smack
15.	PAN	*pakpak	to clap, flap the wings
16.	PWMP	*ragepak	to slap; sound of a collision
17.	PWMP	*ripak	a loud clap
18.	PMP	*rumpak	to thud, strike against
19.	PWMP	*ta(m)pak	to slap, strike with the hand
20.	PMP	*tepak	to slap
21.	PWMP	*tipak	to slap; sound of slapping or clapping
22.	PWMP	*upak	to slap, hit, beat
23.	Iban	cəpa-cəpak	slapping or lapping, as water under a boat
24.	Bontok	dospák	to slap s.o.'s face with the open palm
25.	Javanese	grapak	a branch snapping
26.	Cebuano	gúpak	applause; croaking of frogs, flapping of wings, clacking of wooden slippers
27.	Cebuano	hágpak	to slam with a loud bang
28.	Bare'e	japa	the sound of loud chewing or smacking of the lips
29.	Bare'e	jipa	to clack the tongue as a sign of recalcitrance or reluctance
30.	Sasak	kəpak	to hit, strike
31.	DPB	kurpak	to clap the hands
32.	Bikol	lagapák	sound of a dull smack, a slamming
33.	Keley-i	lalpak	the sound produced when a heavy flat object falls on a hard surface; thud
34.	Maranao	lopak	to club, whip, hit, knock door
35.	Cebuano	lugápak	slapping, cracking sound
36.	Kenyah	mepak	to wash clothes
37.	Rembong	pepak	the sound of a tree branch falling
38.	Rembong	pépak	a bamboo instrument placed in the fields to frighten off rice birds, etc.
39.	DPB	pukpak	sound of many knives cutting at once
40.	Lun Dayeh	rafak	the sound of splashing water
41.	Maranao	rampak	to hit by ricochet; get two birds with one stone
42.	Bikol	sagpák	to bump into s.t.
43.	Bikol	salpák	to collide in mid-air (fighting cocks, kites, airplanes)
44.	Yakan	sampak	to slap
45.	Pangasinan	səpák	to strike with the hand in a quarrel, etc.
46.	Karo Batak	sépak	to kick with the sole of the foot

(continued)

47. Kankanaey	sudpák	to slap in the face; box the ears
48. Bikol	sugpák	to bop; to hit s.o. on the head
49. Bontok	tadpák	to slap
50. Sundanese	talipak	slap or kick with the flat of the hand or foot from a side angle
51. Ilokano	tilpák ~ tiplák	to slap with the open hand
52. Ilokano	tupák	the sound of a soft object falling

*-paŋ	lame	
1. PWMP	*kimpaŋ	lame; to walk with a limp
2. PMP	*timpaŋ	lame, limping; to limp
3. DPB	əppaŋ	to walk with a limp
4. Banggai	kepaŋ	paralyzed, lame
5. Makassarese	keppaŋ	crippled, lame
6. Toba Batak	paŋpaŋ	lame, crippled

*-pap	flattened	
1. PWMP	*apap	flat, level
2. PWMP	*lepap	flattened
3. PWMP	*tepap	to bring down the flat of the hand on s.t.
4. Ilokano	duŋpáp	flattened, crushed, as a hat
5. Ilokano	lugpáp	flat, flattened
6. Keley-i	lukpap	to lie down flat on stomach, face down
7. Ilokano	pappáp	short, flat (said of the nose)
8. Ilokano	tumpáp	flat end, e.g. the bottom of a glass

*-paq	to spit; thing chewed but not swallowed	
1. PWMP	*sapaq	chewed betel quid
2. PWMP	*sepaq	chewed mass of food or betel nut in the mouth
3. PMP	*supaq	to spit, expectorate
4. Ida'an Begak	appaʔ	to chew, masticate
5. DPB	cəpah	s.t. that has been chewed, like betel nut; expectorated betel quid
6. Bikol	kuspáʔ	to spit s.t. out with great force
7. Gaddang	luppa	to spit

(continued)

8. Amis	paʔpaʔ	to chew; the action of jaws and teeth in chewing
9. Amis	sopaʔ	to spit; spittle
10. Tiruray	tumfaʔ	to chew some medicine in place of tobacco in the betel quid and then rub it onto the location of a pain
11. Cebuano	úpaʔ	to chew food for s.o. else, esp. for a child to eat; to spit s.t. out

*-pas		to tear or rip off
1. PWMP	*kupas	to come off in thin layers
2. PWMP	*lepas	to set free, let loose
3. PCEMP	*lupas	to loosen, remove, untie
4. PWMP	*paspas	to remove, take off
5. PWMP	*Ra(m)pas	to tear off
6. POC	*Rupas	to loosen, untie
7. PPH	*tag(e)pás	to cut, chop
8. PWMP	*ta(m)pas	to lop off, cut off
9. PPH	*tiq(e)pas	to cut off with one stroke
10. Banggai	dimpas	to tear oneself loose
11. Dumagat (Casiguran)	gəpas	to cut off the end of a string; trim the loose ends of a mat or basket
12. Buruese	hufa-h	to take off, remove, untie, release
13. Arosi	iha	to unwrap, as food cooked in leaves
14. Wolio	lapa	to escape, get away, be acquitted
15. Bikol	rugpás	to pick all fruit, both green and ripe
16. WBM	tumpas	to finish stripping the mature abaca in a grove
17. Dusun (Kadazan)	upas	to peel betel nut

*-paw		to exceed, surpass
1. PWMP	*la(m)paw	to exceed; excess
2. PPH	*lepaw	to surpass, exceed
3. Ilokano	gimpaw	to surpass, transcend
4. Maranao	limpaw	to go beyond, go higher
5. Ibaloy	on-pawpaw	to surpass others in height
6. Cebuano	úmpaw	to be outclassed or outstripped in comparison

*-pay.₁		to drape over, hang down
1. PWMP	*salampay	to hang or drape over the shoulder, or from a line
2. PAN	*sapay	to drape over the shoulder or from a line, as a cloth
3. PS	*səmpay	to hang over rope
4. Waray-Waray	alampay	kerchief; shawl
5. Iban	ləmpai	to hang down or over, drape
6. Ilokano	salupaypáy	hanging imperfectly, as clothes draped crookedly on the body

Note: Also Amis *caʔfay* 'to drape over the shoulder or shoulders', *cahfay* 'to drape over, of clothing or blankets'.

*-pay.₂		to wave, flap
1. PPH	*payapay	to wave, flap
2. PPH	*paypáy	to wave the hand, as in beckoning s.o. or fanning oneself
3. Malay	ampai	hanging and swaying, of a swaying python, a line for hanging out clothes, tentacles, such as those of the jellyfish, treading water
4. Maranao	kapay	to beckon with hand or fingers
5. Iban	kəpai	to wave, beckon to; flutter
6. Iban	ləpai	to swing the arms (in walking)

*-ped		to block, obstruct; constipated
1. PWMP	*aped	hindrance, obstacle
2. PAN	*pedped	to press together, pack solid
3. Maranao	saliped	to cover, block
4. Puyuma	səpəɖ	constipated
5. Keley-i	siped	to stop or block s.t. in motion, as a goalkeeper blocking a ball

*-peg		to hit, beat (dbl. *-bag, *-beg)
1. PWMP	*pegpeg	to box, give a blow
2. Maranao	parpeg	to hit, beat
3. Javanese	pəg	the sound of a hard blow on the side of the head, as when one's ears are boxed

(continued)

4. Hanunóo	pulpúg	beating, hitting, pounding
5. Balinese	səpəg	to chop violently at, hew
6. Javanese	tampəg	to slap in the face

*-pek		to decay, crumble; sound of breaking (dbl. *-bek₋₂, *-dek, *-mek, *-pak₋₃, *-pik, *puk₋₁)
1. PWMP	*kepek	to flap the wings
2. PAN	*pekpek	to beat, hit
3. PMP	*repek	brittle, friable; to crumble
4. PMP	*tampek	to slap, hit with the open hand
5. PWMP	*upek	to crumble; piece broken off
6. Banggai	dempek	to stutter
7. Nggela	gapo-gapo	to spread the wings; flap, flutter
8. Maranao	gopək	to break, crush
9. Old Javanese	kəḍəpək	thudding or bumping noise
10. Old Javanese	kərəpək	the sound of cracking
11. Kankanaey	kippə́k	to snap, clack (as a cane that breaks)
12. Mapun	lepek	to make a rapid succession of slapping or popping sounds
13. Dumagat (Casiguran)	lipək-lipək	the sound of rain hitting the ground, or the feet of a galloping horse
14. Maranao	lopapək	dust
15. Manggarai	ŋapek	sound of the mouth while eating
16. Karo Batak	rampək	crisp, easily breakable
17. Old Javanese	rimpək	broken up; out of joint, smashed
18. DPB	səmpək	to break off (sugarcane, etc.) close to the ground so as to produce the sound "pek"
19. Toba Batak	silpok	broken up, shattered
20. Balinese	təpək	to slap with the flat hand
21. Toba Batak	tilpok	to break up, shatter
22. Ilokano	topék	to crush, reduce to fine particles

*-pel		plug, stopper; to cram
1. PWMP	*pelpel	to fill, cram full
2. PWMP	*sumpel	stopper, plug, cork
3. PMP	*tampel	to patch, plaster over
4. Sundanese	dumpəl	to plug (a hole), fill up with s.t.
5. Balinese	əmpəl	dam a stream, make a fishpond

(continued)

6. Ilokano	ispél	experiencing difficulty in getting the food down the throat
7. Balinese	kəmpəl	to sit or stand in a throng, be squeezed together
8. Javanese	umpəl	sticking close to each other; jammed together, as passengers in a bus

*-pen	tooth	
1. PMP	*ipen	tooth
2. PAN	*lipen	tooth
3. PAN	*nipen	tooth
4. PAN	*ŋipen	tooth
5. Agutaynen	rigpen	close together, without spaces, as of a bamboo fence, or one's teeth
6. Bikol	sípan	toothbrush (made from a tree branch)

*-peŋ	to plug up, dam; cover (dbl. *-beŋ.₁)	
1. PMP	*qempeŋ	to block, obstruct, dam a stream
2. PWMP	*tampeŋ	to plug or dam up
3. PWMP	*umpeŋ	dam, water barrier
4. Maranao	ampəŋ	to cover, protect
5. Maranao	dapəŋ	to cover
6. Kelabit	əpəŋ	a protective or preventive barrier
7. Lun Dayeh	fefeŋ	bunds built on the paddy field; a dam made across a stream
8. Malay	impaŋ	fish-weir or dammed fish-pond
9. Sika	lepeŋ	to close
10. Maranao	sampəŋ	a cover. lid
11. Maranao	sapəŋ	to cover

*-pes	empty, deflated (dbl. *-pis.₁)	
1. PMP	*i(m)pes	to deflate; subside (of a swelling); empty (rice husk)
2. PWMP	*ke(m)pes	to deflate
3. PMP	*ki(m)pes	to deflate, shrink
4. PMP	*pes	the sound of escaping air, etc.
5. PWMP	*pespes	to squeeze, press out

(continued)

6. PPH	*qepés	deflated, shrunken
7. Manggarai	gimpes	sunken (cheeks), shrunken
8. Manggarai	gipes	shrunken (from crushing pressure)
9. Kambera	kapíhu	flatus; break wind
10. Puyuma	mu-tpes	deflated (balloon, belly of dead animal)

*-pet.$_1$		firefly
1. PMP	*qali-petpet	firefly
2. Bikol	ʔaninipót	firefly
3. Banggai	dipopot	will o' the wisp, ignus fatuus
4. Kiput	kebepet	firefly
5. Ida'an Begak	ləŋipot	firefly
6. Tae'	lumpepeʔ	firefly
7. Melanau (Mukah)	pələpət	firefly

*-pet.$_2$		to hold, grasp, catch
1. PPH	*kapét	to hold onto, cling to
2. PAN	*sapet	to grab with the hand, catch
3. Itbayaten	adpət	support, holding on; holder, that which holds s.t. to another
4. Itbayaten	rapət	holding onto s.t.
5. Itbayaten	yapət	clinging to (as of vines)

*-pet.$_3$		plugged, stopped, closed off
1. PWMP	*ampet	to stanch the flow, as of blood
2. PMP	*ka(m)pet	plugged, stopped, blocked
3. PWMP	*lepet	to plug, stop up
4. PWMP	*pa(m)pet	stopped up
5. PMP	*petpet	sealed, leak-proof
6. PWMP	*rapet	to join together, attach tightly
7. PMP	*sepet	obstructed, of the flow of water
8. PMP	*si(m)pet	to close, shut off
9. Javanese	cumpət	closed firmly
10. Javanese	ḍapət	tightly closed
11. BM	dimpot	to plug or cover up crevices, holes in the roof, etc.
12. Javanese	ḍipət	tightly closed, of eyes

(continued)

13. Balinese	əmpət	to close off, pinch, stuff full, obstruct, shut, lock
14. Sasak	impət	to close, shut
15. Sasak	jəpət	tightly close (as the mouth)
16. Manggarai	kepet	to bind a wound; to dam water
17. Manggarai	képet	to cover with the two hands (genitals, flowing water) to block, close off
18. Paiwan	lʸupetj	plug, stopper
19. Javanese	pumpət	clogged, stopped up
20. Sasak	rəmpət	to block, obstruct (as a path)
21. Sasak	ripət	to close; tight, "hermetically sealed"
22. Balinese	sampət	to block up (a hole), stop the flow of water
23. Balinese	sumpət	stopper, cork

*-pi.$_1$		dream
1. PWMP	*lupi	dream
2. PCEMP	*mepi	dream
3. PMP	*nipi	dream
4. PWMP	*nupi	dream
5. PAN	*Sepi	dream
6. PAN	*Sipi	dream
7. Lampung	hanipi	a dream
8. Tarakan	indupi	to dream
9. Wolio	kaɲipiɲipi	to indulge in daydreams, muse, constantly think of s.t.
10. Kayan (Uma Juman)	ñupi	dream
11. Kelabit	upih	a dream
12. Gitua	vivi	to dream

*-pi.$_2$		to fold (dbl. *-piq)
1. PPH	*lapi	to fold over
2. PWMP	*lumpi	a fold, hem
3. Iban	ləpiʔ	fold, hem; to double over
4. Keley-i	lugpi	to fold s.t., as cloth or paper
5. Iban	rəpiʔ	(of cane, grass, twig, etc.) bent, kinked, bent back on itself
6. Amis	ʔəpi	hem

*-pid		to braid, wind together (dbl. *-bid)	
1.	PWMP	*epid	to braid, intertwine
2.	PPH	*la(m)pid	to braid, intertwine strands
3.	PAN	*qapid	to braid, twine
4.	PAN	*qupid	a braid; to braid
5.	PWMP	*salapid	a braid
6.	Kelabit	belipid	intertwining
7.	Kankanaey	ípid	thatch wall
8.	Bikol	ma-lipíd-lipíd	twisted (as a tree branch)
9.	Sundanese	rampid	to wind thread around a spool
10.	Mandaya	sapid	braid
11.	Cebuano	sulápid	to braid, plait; walk crossing the feet over each other

*-pik		to pat, slap lightly (dbl. *-pak₋₃, *-pek, *puk₋₁)	
1.	PWMP	*ampik	to pat, clap
2.	PMP	*gepik	sound of light smacking
3.	PMP	*lepik	to snap, break off (twigs, etc.)
4.	PWMP	*pik	to click
5.	PAN	*pikpik	to pat, tap lightly
6.	PWMP	*Repik	to break to pieces (s.t. brittle)
7.	PMP	*ta(m)pik	to slap, pat lightly
8.	PAN	*tepik	a pat, light slap
9.	Cebuano	hágpik	to slap s.o. fairly hard on the shoulders
10.	Malay	kələpek	flip-flop; repeated flips
11.	Sika	kepik	wing, fin
12.	Kankanaey	kippík	to snap, crack (as a breaking twig)
13.	Agutaynen	lapik	to intentionally snap or break s.t. in half, or break s.t. off
14.	Maranao	latpik	crackling sound
15.	Kankanaey	lipíik	to crack
16.	Mandaya	pagikpik	wing
17.	Bikol	parikpík	fin, flipper
18.	Kankanaey	pikípik	to drop, trickle, leak, drip, dribble
19.	Fijian	sabi	to slap or hit with the open hand, not so strong as saba-ka
20.	Tongan	sipi	to smack, slap, hit with the palm of the hand
21.	Bontok	tadpík	to slap lightly, usually with the flat of the palm

Note: Also PWMP *bikbik 'sound of slapping, bubbling, etc.'

*-pil		to attach, join
1. PMP	*ha(m)pil	to go together
2. PWMP	*tampil	to attach, connect, join
3. PWMP	*tumpil	to attach or join
4. Cebuano	apíl	to be included
5. Malay	dampil	touching one another (of contact, in contrast to mere proximity)
6. Karo Batak	gumpil	to form a cluster; swarm of bees, etc.
7. Karo Batak	rumpil	to romp with one another
8. Karo Batak	umpil	to do s.t. together, with concerted action

Note: Also PWMP *tampir 'to attach, connect, join'.

*-pin		protective layer, as of cloth or clothing; sleeping mat
1. PMP	*hapin	liner, layer, insulation, padding; sleeping mat
2. PWMP	*lampin	layer of cloth or clothing; diaper
3. PPH	*sapín	lining, insulation, padding; underlayer, as of clothing
4. Kelabit	əpin	sleeping mat
5. Lun Dayeh	təfin	diaper

Note: Also Kayan *tebin* 'a layer (as of clothing)'. English (1986) gives Tagalog *sapín* as a borrowing of Spanish *chapin*, a form that I am unable to find in standard Spanish-English, English-Spanish dictionaries, and which this comparative evidence shows is almost certainly native.

*-piŋ		next to, beside
1. PPH	*sípiŋ	next to, in close proximity
2. Tae'	ampiŋ	near, close by
3. Aklanon	ípiŋ	to be side-by-side with, next to
4. Malay	sampiŋ	side, flank, border
	di-sampiŋ	beside, next to

*-piq₋₁		to break off a piece
1. PWMP	*Rapiq	to crumble, break s.t. to pieces
2. PAN	*sepiq	to break off fruit or branches
3. PPH	*sipiq	to break off a component piece of s.t., as twigs or branches from a tree

(continued)

4. Chamorro	famfeʔ	to pick fruit, break fruit off a tree; pull a tooth
5. Balinese	kəpih	to crumble, break away in bits
6. Malay	rəpeh	crumbling; to pick to pieces
7. BM	tapiʔ	breaking off of branches
8. Chamorro	tifeʔ	to pick, break off, esp. the fruit from trees, vines, etc.; pull a tooth

Note: Also Binukid *bagtiʔ* 'to break, snap off (s.t. brittle or crisp)'.

*-piq₋₂	to fold (dbl. *-pi)	
1. PWMP	*gepi(q)	to fold, folded
2. PWMP	*kepiq	to fold, folded
3. PWMP	*lepiq	a fold, hem
4. PWMP	*lupiq	a fold, hem
5. PPH	*tupíq	to fold, crease
6. Muna	api	to make twofold, double
7. Ida'an Begak	kippiʔ	folded cloth
8. Ilokano	kulpí	a fold
9. Tagalog	kupîʔ	folded, doubled
10. Kankanaey	lokpí	to fold, fold up
11. Keley-i	lugpi	to fold s.t., as cloth or paper
12. Balinese	tampih	to fold up, put in layers
13. Fijian	tibi	to bend sharply, fold clothes; plait

*-pis₋₁	to deflate; empty (dbl. *-pes)	
1. PWMP	*e(m)pis	empty rice husk
2. PMP	*kampis	to deflate, shrink
3. PWMP	*ke(m)pis	to shrink, shrivel
4. PWMP	*pis	to hiss
5. PWMP	*umpis	deflated, empty
6. Kankanaey	apípis	to lessen, reduce, diminish
7. Ngaju Dayak	apis	hollow, empty grains (of rice)
8. Karo Batak	gəmpis	empty, flat
9. Maranao	kimpis	to shrivel
10. Tagalog	kumpís	deflated (reduced from swelling or bulge)
11. Tagalog	ʔumpís	deflated (said of tires, balloons, etc.)
12. Isneg	lappít	empty heads of rice

Note: The Tagalog forms cited here may reflect *-pes.

*-pis.₂		thin, of materials
1. PAN	*iŋepis	thin (of materials)
2. PMP	*lapis	stone slab, thin layer
3. PWMP	*lipis	thinness of materials
4. PWMP	*lupis	thinness of materials
5. PMP	*mepis	thin
6. PMP	*mipis	thin, of materials
7. PAN	*Nipis	thinness (of materials)
8. PWMP	*pispis	thin, of materials
9. PWMP	*Rampis	thin, thin layer
10. PWMP	*taRpis	thinness, fineness of texture
11. PMP	*tipis	thin (of materials)
12. Balinese	ampis	thin, worn thin with use
13. Manggarai	dempis	thin and flat
14. Dusun (Kadazan)	dipis	thin
15. Bikol	hiŋpís ~ himpís	thin (things)
16. Dusun (Kadazan)	kampis	flat and thin
17. Amis	kohpic	thin (opp. thick)
18. Paiwan (Western)	lʸusepit	thin (as paper), thin, as a person
19. Bare'e	sampi	to become thin or narrow
20. Sangir	sipiʔ	thin
21. Ivatan	taripis	thin (objects)
22. Hiligaynon	yágpis	lanky, thin

*-pit.₁		near, close to
1. PWMP	*dapit	near, close to
2. PAN	*ipit	near; come near; edge, border
3. Mansaka	ăpit	near, close to, at the side of
4. Masbatenyo	halapít	near
5. Kapampangan	lápit	to draw near to s.t.
6. Mapun	tapit	nearness, proximity

Note: Apparently distinct form *-pit.₂.

*-pit.₂ (or *piqit?)		to press, squeeze together; narrow
1. PWMP	*ca(R)pit	pincers, tongs
2. PWMP	*cupit	narrow
3. PWMP	*diq(e)pit	to join, fasten together lengthwise
4. PMP	*ga(m)pit	to hold together
5. PMP	*gepit	to pinch, squeeze between
6. PMP	*gipit	narrow, tight, confined
7. PWMP	*gu(m)pit	to pinch together; narrow

(continued)

8.	PWMP	*he(m)pit	to press between two surfaces
9.	PPH	*hig(e)pít	to pinch or squeeze between two surfaces
10.	PAN	*kapit	to fasten together by sewing or tying
11.	PAN	*kepit	to press together; pressed together
12.	PAN	*kipit	narrow; to pinch between tongs, etc.
13.	PAN	*kupit	to press together, close tightly
14.	PWMP	*lapit	to put close together
15.	PMP	*le(m)pit	to fold
16.	PPH	*lig(e)pít	to fasten, as with a clip
17.	PMP	*liqepit	to press between two flat surfaces
18.	PMP	*luqepit	to press between two flat surfaces
19.	PWMP	*piqit	closed tightly, as the eyes
20.	PAN	*pitpit	to pinch, squeeze, press between two surfaces
21.	PMP	*qapit	tongs, anything used to hold things together by pinching
22.	PAN	*qepit	to squeeze, press between
23.	PAN	*qipit	pincer of crustaceans; tongs; to press together, pinch, squeeze
24.	PPH	*qiR(e)pit	to pinch between two surfaces, clip
25.	PWMP	*rupit	narrow
26.	PMP	*Ri(m)pit	to squeeze between; pinch; take between fingers
27.	PWMP	*sapit	to clamp, pinch together
28.	PWMP	*se(m)pit	narrow; squeeze tightly
29.	PPH	*sig(e)pít	to clip on
30.	PAN	*sipit	tongs; claw of a crab or lobster
31.	PPH	*siR(e)pít	a clip used to fasten things together
32.	PWMP	*supit	narrow; to pinch; tongs
33.	PAN	*Sapit	to hold together
34.	PWMP	*upit	to press, squeeze together
35.	PWMP	*zapit	to pinch, squeeze, press
36.	PWMP	*zepit	to pinch, take up with the fingers
37.	Kankanaey	agpít	to carry under (one's arm, etc.)
38.	Kankanaey	alípit	to carry under (one's arm, etc.)
39.	Banggai	ampit	what is folded, a fold, pleat
40.	Malay	mən-cəpit	to nip
41.	Ilokano	daípit	to crowd, throng; press, stick together
42.	Banggai	dapit	to clip, pinch
43.	Malay	dəmpit	pressed together; in contact
44.	Keley-i	ʔəgpit	to place things close together; to close the armpit in order to hold s.t. there
45.	Kambera	haŋgápitu	tongs
46.	Keley-i	həlpit	to put things closely side by side

(continued)

47. Kapampangan	igpit	to tighten s.t.
48. Javanese	jimpit	a pinch
49. Kambera	júmbitu	to pinch
50. Iban	jumpit	tight (of clothes)
51. Iban	kəripit	drawn together (as the sides of a coat one is wearing)
52. Kenyah	ketepit	fused together
53. Ifugaw	kuhípit	narrowness of things
54. Mapun	kulipit	slanted (of the eyes)
55. Hiligaynon	lágpit	rat trap
56. Kambera	ŋápitu	to clip, pinch
57. Manggarai	ŋgepit	pincers; to nip, catch between pincers
58. Bontok	pipʔít	to be crowded, have no room
59. Amis	ʔalapit	tweezers, chopsticks
60. Ma'anyan	rampit	to tighten s.t. that is loose
61. Lun Dayeh	refit	a hair clip
62. Bikol	sagipít	a clip; pincers; tongs
63. Isneg	sagpít	the tongs of the blacksmith
64. Dumagat (Casiguran)	səlpet	pincers, tongs, clothespin; to pinch (as finger in door)
65. Balinese	sripit	narrow (road)
66. Bikol	sugpít	to harvest rice by pulling the heads of grain off between the fingers
67. Hanunóo	tipít	clamps, holders, as used to fasten house walls down
68. Ifugaw	ukpít	to hold by pressing s.t. between the knees, or between the elbow (or upper arm), and the side of the body
69. Gorontalo	wepito	narrow
70. Bikol	yaʔpít	narrow (as a passage), tight (as clothes)

Note 1: Longer variants of -CVC roots normally are suggested because of the appearance of an otherwise unexpected glottal stop within a -CVC root in Philippine languages, which does not appear to reflect *q. This root differs from others in showing the reflex of *q in Malayic languages, as in Malay (Brunei) *pihit* 'to press down; to weigh down', Kadayan *pihit* 'to pinch' which, together with Casiguran Dumagat *piʔít* 'tight, narrow, crowded', Tagalog *piʔít* 'squeezed, pressed tightly between two persons or things', Banggai *piit* 'shut, of the eyes', supports PWMP *piqit. It also differs in being one of a handful of non-onomatopoetic roots that are attested as free-standing.
Note 2: Source for Ma'anyan *rampit* unknown.

*-puk.$_1$		to thud, snap, crack, break (dbl. *-pak.$_3$, *-pik, *-pek)
1. PMP	*cepuk	dull sound

(continued)

2. PAN	*Cepuk	to hit, sound of hitting, clapping, thumping,	
3. PMP	*depuk	to snap, crack	
4. PMP	*epuk	to make a popping sound	
5. PMP	*gepuk	to beat, break to pieces	
6. PMP	*kepuk	to beat, crunch, break	
7. PWMP	*lepuk	to fall with a thud	
8. PPH	*lipuk	to snap, break off	
9. PMP	*puk	pop! plop! splash!	
10. PAN	*pukpuk	to hammer, pound, beat	
11. PMP	*rimpuk	to strike together	
12. PMP	*rupuk	the sound of breaking or cracking	
13. PMP	*Rimpuk	to break with a loud sound	
14. PWMP	*sa(m)puk	to collide, bump into	
15. PWMP	*sempuk	to bump or bang into	
16. PMP	*tampuk	to clap, thud	
17. PWMP	*ti(m)puk	to clap, clack	
18. PMP	*upuk	to clap, slap, pound	
19. Manggarai	cépuk	the sound of tapping, breaking, etc.	
20. Palawano	dumpuk	to hit, collide, bump	
21. Karo Batak	gupuk	to come to blows	
22. Manggarai	japuk	to crunch, chew noisily	
23. Manggarai	jipuk	the sound of a pig when eating	
24. Old Javanese	kərəpuk	the sound of clashing, clanging, clattering	
25. Agutaynen	lagopok	a loud, hollow, smacking sound caused by one object hitting against another	
26. Agutaynen	lopok	for s.t. to burst; for s.t. to explode with a loud popping sound	
27. Bikol	lugpók	to hit s.o. with s.t. clasped in a closed fist	
28. Karo Batak	nampuk	to fall down, of ripe fruits, as the durian	
29. Manggarai	ŋapok	the sound of a slap	
30. Manggarai	ŋapuk	the sound of chewing	
31. Amis	parokpok	to gallop	
32. Lun Dayeh	rafuk	clashing together, as in a soccer game	
33. Manggarai	rapok	to slap	
34. Mandar	reppo?	to break (as a collapsing house floor)	
35. Bikol	sagpók	to knock two heads together	
36. Dumagat (Casiguran)	salpók	to beat against, to slap, to collide; impact	
37. DPB	səlpuk	to snap or break in the middle (as young bamboo stalks)	
38. Maranao	simpok	to touch lightly, hit lightly	

(continued)

39. Palawano	talpuk	wham! Sound of s.t. heavy hitting the ground
40. Ibaloy	tanipok	to slap s.o.
41. Ida'an Begak	tədupuk	to collide
42. Paiwan	tsupuk	to hit with a club
43. Ida'an Begak	ukpuk	to hit with a hammer

*-puk.$_2$	dust (dbl. *buk.$_1$)	
1. PMP	*apuk	dust
2. PMP	*lapuk	decayed, rotten, moldy (of wood)
3. PMP	*repuk	fragile, brittle, easily broken (of wood)
4. PWMP	*rupuk	rotten, crumbling, of wood
5. PMP	*Re(m)puk	crumbling, of wood
6. PMP	*sapuk	fine dust, airborne dust
7. PPH	*tapuk	dust in the air; to stir up dust
8. Ibaloy	depok	dust
9. Bikol	dupók	rotten (referring to wood)
10. Tagalog	gapók	rotten inside, hollowed out by weevils or termites (said of wood)
11. Ifugaw	húpuk	dust, sand
12. Agutaynen	kolapok	dust, dusty
13. Fijian	kuvu	dust, spray, smoke
14. Tboli	lufuk	floating impurities in water or wine
15. Kenyah	mopok	rotten (of wood)
16. Bimanese	numpu	to pulverize
17. Pangasinan	pokpók	to shake off dust
18. Tetun	rahuk	brittle, fragile
19. Balinese	səpuk	dust
20. Ayta Abellen	toapok	dust, dust-like soil; dusty
21. Agutaynen	topok	brittle, dried out, as of old bamboo; become worn out, frayed, brittle, or rotten

*-puk.$_3$	to gather, flock together	
1. PMP	*lumpuk	a cluster, group
2. PWMP	*simpuk	to heap up, make a pile
3. PWMP	*ta(m)puk	heap, pile
4. PWMP	*tumpuk	a small heap, rounded mass
5. PWMP	*umpuk	a heap, pile, collection
6. Arosi	ahu	to flock together

(continued)

7. Bunun	isʔampuk	gathering; assembly
8. Malay	kəlompok	cluster, group
9. Iban	kəmpok	moulded, squeezed into shape (as balls of sago)
10. Ifugaw	pukúpuk	to make a small heap by pouring out pounded rice, beans, peas or the like
11. Dumagat (Casiguran)	talumpúk	a pile, mount of s.t. piled up
12. Ilokano	tarapók	to gather, assemble, come together, flock together, associate
13. Javanese	təmpuk	to flow together, join
14. DPB	tərpuk	cluster, group

*-pul		blunt, dull
1. PMP	*de(m)pul	blunt, dull
2. PMP	*dumpul	blunt, dull
3. PPH	*pulpul	blunt
4. PMP	*tumpul	blunt, dull
5. Malay (Jakarta)	puntul	blunt
6. Tiruray	təmful	dull, not sharp
7. Makassarese	tippuluʔ	dull, blunt

*-pun		to assemble, collect, gather, heap up (dbl. *-bun.$_2$)
1. PPH	*ampun	to gather, collect together
2. PWMP	*e(m)pun	to gather, assemble
3. PWMP	*punpun	to assemble, gather
4. PMP	*qi(m)pun	heap, collection; to gather, heap up
5. PWMP	*ra(m)pun	to gather, collect, put in one place
6. PMP	*rempun	to gather together
7. PWMP	*ri(m)pun	to collect, gather
8. PWMP	*rumpun	cluster, clump
9. PAN	*tapun	to gather, assemble
10. PWMP	*tipun	to assemble, come together
11. PMP	*upun	to gather in masses, as smoke
12. Itbayaten	adpon	pile, gathered leftover or mound made of unburned trees and branches after burning the field
13. WBM	hurapun	to alight in a flock (birds)
14. BM	kapun	to meet, assemble
15. Buruese	lupu-k	to assemble
16. Dumagat (Casiguran)	paluŋpun	to stack up rice stalks after harvesting

(continued)

17. Kavalan	sapun	to gather, collect
18. Tontemboan	sipun	to gather
19. Kankanaey	sugpún	to put their money together; enter, partnership with somebody
20. Ida'an Begak	təripun	to gather
21. Ngadha	tupu	to gather, assemble, collect

*-puŋ-1		a bunch, cluster
1. PMP	*kampuŋ	assembly, meeting
2. PWMP	*lumpuŋ	a cluster, group
3. PMP	*puŋpuŋ	bunch, cluster, as of fruit
4. PAN	*qupuŋ	a bunch, cluster
5. PMP	*rimpuŋ	to tie the feet together
6. PWMP	*tampuŋ	to collect, gather
7. Balinese	ampuŋ	to harvest rice
8. Ayta Abellen	dopoŋ	to gather together in groups (of people)
9. Iban	əmpoŋ	to collect, call or tie up together
10. Iban	gəmpoŋ	gathering, collection, bundle, sheaf; to assumble, call together
11. Iban	gərəmpoŋ	a bundle or sheaf
12. Hiligaynon	húgpuŋ	a bundle, bouquet
13. Yami	ipoŋ	a bunch, unit of fruit (longan, grapes, wax apples, etc.)
14. Ilokano	lípoŋ	a crowd of people
15. Lamaholot	mupũ	group, flock
16. Malay	rəmpoŋ	a bunch; make into a bunch
17. WBM	rupuŋ	a hand of bananas
18. Puyuma	sarəpuŋ	to gather
19. Ilokano	tarapóŋ	to join, come together, unite, gather
20. Toba Batak	tupuŋ	a rolled-up leaf; gather, assemble

*-puŋ-2		to float
1. PMP	*apuŋ	to float
2. PWMP	*gampuŋ	to float
3. PWMP	*lepu-lepuŋ	floating
4. Malay	tər-kapoŋ kapoŋ	bobbing up and down on the surface of water, as driftwood or a floating leaf
5. Iban	lapa-lapoŋ	Floating
6. Kelabit	l-em-upuŋ	to float
7. Malay	pəlampoŋ	floating mark; float of line or net; impromptu raft

*-puŋ.3		rice flour
1. PWMP	*tapuŋ	rice flour
2. Rejang	gəlpuŋ	flour, meal
3. Muna	golupu ~ gulupu	fine flour from rice, maize or cassava, made to form a coating on fried bananas, fried yams, etc.
4. Malay	təpoŋ	flour, meal

Note: Also Bolaang Mongondow *topoŋ* 'flour'.

*-puq		brittle
1. PWMP	*Rapuq	rotten, crumbling
2. PWMP	*tepuq	brittle
3. Maranao	gəpoʔ	to break, as branches or twigs
4. Cebuano	hágpuʔ	for a line or rope to break under tension
5. Manggarai	mepu	brittle, fragile; decayed

*-pur		to mix (dbl. *-bur, *-buR, *-huR)
1. PMP	*ca(m)pur	to mix, blend
2. Ilokano	gampur-an	to mix, blend
3. Manggarai	jampur	to mix; together with
4. Ilokano	kampór	mixture, medley

*-pus.1		to end, finish; used up
1. PMP	*epus	stump, stub; to finish, complete
2. PPH	*kapus	to run out of supplies
3. PMP	*ke(m)pus	to come to an end
4. PWMP	*la(m)pus	gone, vanished, terminated
5. PMP	*le(m)pus	finished, used up, gone
6. PMP	*mampus	to use; used up
7. PWMP	*puspus	finished, completed; all gone
8. PMP	*qa(m)pus	to come to an end, be destroyed
9. PAN	*tapus	to end, complete, finish
10. PMP	*te(m)pus	to end, conclude
11. PWMP	*timpus	to finish, use up
12. PMP	*tu(m)pus	to complete, finish
13. PMP	*upus	to end, finish; finished, used up
14. Kankanaey	dokpús	to finish, complete, terminate, end; to perfect

(continued)

15. Rembong	dupus	to finish
16. Tagalog	gipós	reduced to a stub; extinguished
17. Hiligaynon	hiŋapús	to end, finish, complete, terminate
18. Kambera	hupu	to end, finish
19. Maranao	ipos	done with, gone away
20. Lakalai	kaluvu	to come to an end; finished
21. Kankanaey	kipús	quickly exhausted; up in a short time
22. Kankanaey	kudpús	up; consumed, wasted, finished
23. Isneg	laŋpút	to finish, complete
24. Mansaka	lipos	small remaining portion (as of soap); butt (from a cigarette); residue (as ashes of s.t. burned)
25. Kankanaey	luŋpús	finished, completed, perfected
26. Manggarai	mempos	stump, fag-end of anything
27. Kankanaey	ŋopús	brand; piece of wood whose largest part has been burnt
28. Wolio	ropu	to finish off, take all for oneself
29. Iban	rupus	all over, completely
30. Buruese	sepu	finished, completed, done, exhausted
31. Ilokano	talipupús	to do, etc. completely, entirely, utterly (extinguishing a debt, a lineage by death, buying up a crop; devastating a whole village, etc.)
32. Kankanaey	toŋpús	finished, completed, concluded
33. Ilokano	torpús	to finish, conclude, end, terminate

*-pus₋₂		the sound of escaping air
1. PWMP	*lepus	to puff, blow into
2. PMP	*pus	a hiss, whizzing sound
3. Malay	dəmpus	to puff in one's sleep
4. Malay	dəpus	the sound of a rush of air through a narrow opening, e.g. from bellows
5. Malay	əmpus	to blow
6. Javanese	kəpus	to blow on, exhale strongly
7. Kankanaey	kippús	to break wind noiselessly

*-put		to puff, blow hard
1. PWMP	*eput	to puff, blow suddenly
2. PMP	*leput	to blow out with force
3. PWMP	*put	puffing sound

(continued)

4. PMP	*putput	to puff, blow, expel air rapidly, as in using a blowgun
5. PWMP	*seput	blowpipe, blowgun
6. POC	*upu	to blow with the mouth
7. Kenyah	keliput	blowpipe
8. Ilokano	lípot	repeated sound of breaking wind
9. Kenyah	peliput	blowpipe
10. Javanese	sruput	sipping or sucking sounds
11. Cebuano	súlput	to issue forth suddenly, pop out of

The *q- roots

*-qit		hook, barb	
1. PMP	*kaqit	hook-shaped	
2. PAN	*raqit	a hook-like structure	
3. PAN	*saqit	a hook; to hang on a hook	
4. PWMP	*siqit	barb, hook	
5. Kanowit	it	barb	
6. Sasak	kəlait	to hang s.t. on a hook	
7. Malay	ruit	bent; hook-shaped; hook or barb	

The *r-roots

*-raŋ		bright, of light	(dbl. *-daŋ, *-naŋ, *-laŋ)
1. PWMP	*teraŋ	clear, bright	
2. Tae'	arraŋ	light; to shine; glimmer	
3. Malay	cəraŋ	patch of forest where the foliage is relatively thin, allowing light to shine through	
4. Malay	bər-dəraŋ siaŋ	early morning, about 5:30 A.M. (arrival of light)	
5. Puyuma	səraŋ	very bright; sun (in ritual contexts)	

*-riC		the sound of ripping, etc.	
1. PWMP	*berit	to split, tear open	

(continued)

2. PAN	*biriC	to tear, rip
3. PMP	*cirit	the sound of squirting
4. PAN	*geriC	the sound of ripping, gnawing, etc.
5. PWMP	*zerit	shrill or crackling sound
6. Malay	cəŋkərit	field-cricket: *Brachytrypes [sic] portentosus*
7. Malay	cərit	to squirt out in small quantity, e.g. in defecation
8. Tontemboan	gorit	to saw
9. Paiwan	gurits	to squeal (pig)
10. Sundanese	hərit	eerie rustle of the wind, as through bamboo leaves
11. Ngadha	iri	tearing, sound of tearing
12. Cebuano	kádlit	to strike a match
13. Malay	kərit	the sound of scratching, grating or gnawing; to gnaw
14. Itbayaten	pirit	tear, torn apart (of clothes, paper, etc.)
15. Puyuma	siriṭ	to tear
16. Tetun	surit	to have diarrhoea

*-rik		a spot, freckle
1. PMP	*barik	striped, streaked
2. PWMP	*burik	speckled
3. PWMP	*curik	spotted, speckled
4. PBT	*kori	skin disease (ringworm; light spots; tinea, etc.)
5. PMP	*turik	to mark with a line or spot
6. Madurese	balurik	spotted
7. Javanese	blirik	speckled, flecked with color
8. Iban	jurik	striped
9. Malay	kurek	speckled, of a fowl's markings
10. Malay	lorek	very delicate markings (of certain markings on snakes, markings seen in certain lights only, scratches on a polished surface, etc.)
11. Tetun	makerik	colorful; dappled, of a horse
12. Iban	surik	stripe

Note: Also Tiruray *barək* 'the spots on an eel', Kenyah *turek* 'spotted'.

*-ris.₁		rustling sound (doublet *-rus.₁)
1. PPH	*bug(e)ris	to flow through a small opening, of rushing water
2. PWMP	*risris	light rustling sound
3. Sangir	balehiseʔ	rustling, as of wind
4. Iban	bəris	shower, drizzle, spray
5. WBM	biris-biris	of rain, to sprinkle or fall lightly
6. Ilokano	burís	have diarrhoea, flux
7. Bahasa Indonesia	deris	a sound like that of walking through dry grass, cloth being cut, etc.
8. Sundanese	gəbris	the sound of sneezing, etc.
9. Toba Batak	heres	to rustle, crackle
10. Pangasinan	larís	the sound of whipping or lashing
11. Tombonuwo	momoris	to drizzle
12. Kambera	ŋirihu	the sound of scratching, as of a file on something
13. Malay	ris	a rustling sound
14. Kayan	serih	the rustling sound of a rapidly moving animal
15. Wolio	siri	to rustle, murmur

*-ris.₂		to scratch a line (dbl. *-rit)
1. PWMP	*bad(e)ris	draw a line
2. PAN	*Curis	to scratch a line
3. PMP	*garis	to scratch; draw a line
4. PWMP	*guris	a line; to scratch a line
5. PMP	*hiris	a slice
6. PMP	*karis	to scratch, scrape
7. PWMP	*kuris	to scratch, mark with a line
8. PWMP	*Ruris	to tear into strips
9. PMP	*taris	to scratch a line
10. PAN	*turis	line, stripe; to scratch a line
11. Balinese	baris	a strip, row, rank
12. Sangir	behiseʔ	a line, stripe
13. Wolio	bori	a line, borderline
14. Cebuano	búdlis	a long thin mark, stripe, smear
15. Wolio	buri	to write
16. Lamaholot	kənaris	a line, dash
17. Keley-i	kuʔlih	a mark or writing made with a writing instrument
18. Madurese	liris	striped, of batik
19. Puyuma	paris	to scratch inadvertently (as on passing a plant with barbs)

(continued)

20. Puyuma	taʔuris	to scratch, leave a scratch mark
21. Ngadha	teri	a line, stroke, incision
22. Maranao	tiris	a stripe
23. Kankanaey	ugális	striped, streaked
24. Bahasa Indonesia	uris	to scratch; draw a line

Note: Also PWMP *kuRis 'to stratch, mark with a line'.

*-ris.$_3$		a slice; to slice
1. PWMP	*biris	a slice; to slice
2. PMP	*giris	to cut or slice
3. PMP	*qiris	to cut, slice
4. Tagalog	hílis	a cut or sliced portion shaped like a parallelogram
5. Bintulu	ruris	a slice

*-rit		to scratch a line (dbl. *-ris.$_2$)
1. PWMP	*berit	to scratch a line
2. PMP	*burit	a line, stripe
3. PWMP	*curit	a mark, line, dash
4. PCEMP	*erit	to scratch, scrape
5. PWMP	*garit	striped, having stripes or streaks of different color
6. PWMP	*gurit	to scratch; the sound of scratching
7. PMP	*kerit	to scratch or grate
8. PMP	*kurit	a scratch mark, line
9. PWMP	*parit	to mark with lines
10. PMP	*qurit	stroke, stripe, line
11. Kelabit	arit	a design
12. Aklanon	bádlit	lines or marks in the palm
13. Mansaka	baglit	to mark; to draw (a line)
14. Kelabit	barit	striped, colorful
15. Ilokano	bírit	a scar or scarlike indentation on the eyelid
16. Malagasy	fáritra	a mark, an outline, a defined boundary
17. Toba Batak	gorit	a scratch, line
18. Kayan	harit	to cut or scratch
19. Iban	jərit	a line standing out from the ground in woven material
20. Aklanon	kádlit	to mark, draw a line

(continued)

21. Isneg	karít	line, on paper, etc.
22. Selaru	kérit	to scratch
23. Hiligaynon	kúdlit	a line, mark
24. Ilokano	okrít	to line, mark with a line or lines, to scarify, to scratch or cut the skin
25. Ilokano	orárit	streaked
26. Gorontalo	pilito	a line; to mark with a line
27. Ngadha	teri	a line, stroke, incision

Note: Gorontalo *pilito* may contain a reflex of *-ris-$_2$ 'scratch a line'.

*-rud		to scratch, scrape (dbl. *-Rud)
1. PBT	*geru	to scratch
2. PPH	*hírud	to scrape, rub off
3. PBT	*keru	to scrape, grate
4. PWMP	*kerud	to scrape, grate
5. PWMP	*kurud	to scrape
6. PMP	*parud	a rasp, file
7. PWMP	*rudrud	to scrape, rub
8. Balinese	bərud	to scratch, graze, peeling of skin
9. Malay	dərut	a dull scraping sound
10. Malay	garut	scraping against one another
11. Malay	gərut	any dull scraping sound
12. Malay	karut	a rasp
13. Sasak	sərut	to plane, shave off with a plane
14. Hanunóo	tarúd	scraping, rubbing

Note: Bikol *súrod* 'a small-toothed comb used for removing lice' reflects *sujud.

*-rus-$_1$		rustling sound (dbl. *ris-$_1$)
1. PWMP	*garus	to scratch
2. PWMP	*kerus	a rustling sound, as of leaves in the wind
3. PWMP	*rusrus	a rustling sound
4. WBM	dəgurus	a swishing sound of rain falling or water flowing
5. Malay	dərus	a dull rustling or scratching sound
6. Balinese	parus	to stream out strongly; (of water), to stream out when a dam is opened, babble

(continued)

7. Kayan	seruh	the sound of an object dropping through leaves of trees
8. Maranao	təros	to spurt, drip, ooze

*-rus.₂		to slip or slide off (dbl. *-lus)
1. PWMP	*burus	to slip off or away
2. PWMP	*durus	to slide down
3. PWMP	*hurus	to slip or slide down or along
4. PWMP	*lurus	to slip off
5. PAN	*perus	to slip or slide off
6. PWMP	*rusrus	to slip down or off
7. Puyuma	darus	slippery because of mud
8. Tetun	dorus	to slide
9. Simalur	forus	to slip, slide (as down a coconut tree when climbing it)
10. Sasak	gərus	to slip through somewhere

Note: Also Puyuma (Tamalakaw) *mu-HruRus* 'slipped off'.

The *R-roots

*-Raŋ		clanging sound (dbl. *-Reŋ, *-Riŋ, *-Ruŋ)
1. PMP	*keRaŋ	a deep reverberating sound
2. Chamorro	aʔgaŋ	loud noise; booming, resounding
3. Malay	bəraŋ	to clang
4. Malay	chəroŋ-chəraŋ	to clang, as anklets when their wearer moves about, or as bar-iron when being taken to a foundry
5. Malay	dəraŋ	clanging sound, as of a gong, the tinkle of pieces of jewellery, the chink of dollars, the sound of the royal drums
6. DPB	diŋgəraŋ	k.o. drum
7. DPB	gəndəraŋ	drum with skin covering on one end and wood on the other

*-Raq	red	
1. PWMP	*baRaq	red
2. PMP	*daRaq	blood
3. PMP	*ma-iRaq	red
4. PMP	*siRaq	reddish, color of flames
5. Chamorro	agagaʔ	red
6. Chamorro	ataggaʔ	reddish
7. Aklanon	eagáʔ	red, reddish
8. Binukid	lígaʔ	red
9. Manobo (KC)	me-lalegaʔ	red
10. Kavalan	tbaRiʔ	red

*-Raw	hoarse	
1. PWMP	*gaRaw	hoarse
2. PWMP	*paRaw	hoarse
3. PWMP	*peRaw	hoarse
4. WBM	ragəw	to cause hoarseness in the voice

*-ReC	to tighten, as a belt	
1. PWMP	*beRet	belt
2. PPH	*hiRét	to tighten; constriction
3. PAN	*SeRet	to bind tightly; belt
4. PWMP	*zeRet	noose trap
5. PWMP	*ziRet	noose trap
6. Maranao	gagət	tightness
7. Lun Dayeh	giret	belt
8. Tboli	hilet	belt
9. Keley-i	kiʔlət	to tighten s.t., as noose, belt, G-string, or cloth for carrying a baby
10. Pazeh	m-axes	tight (of a belt)
11. Iban	sirat	man's waist or loin cloth

*-Reŋ	to groan, moan, snore (dbl. *-Raŋ, *-Riŋ, *-Ruŋ)	
1. PMP	*beReŋ	a deep grunt or groan
2. PMP	*deReŋ	the sound of groaning or moaning
3. PMP	*heReŋ	to groan, moan, roar, growl
4. PWMP	*legeŋ	a boom, thundering sound
5. Roviana	baroɲo	to snore

(continued)

6. Lun Dayeh	dureŋ	soft speech, a murmur; a complaint, a grumble
7. Kelabit	gareŋ	roaring of an animal
8. Old Javanese	gərəŋ	to growl, snarl
9. Aklanon	hágoŋ	to groan, moan
10. Manggarai	jereŋ	an interjection of disappointment
11. Amis	kaləŋ	a deep voice
12. WBM	lageŋ	the human voice
13. Itbayaten	mayəŋ	of the sound *yeŋ*, of high-pitched tone
14. Kavalan	q<um>Reŋ	roaring (of breakers)
15. Kavalan	ReReŋ	to make a threatening sound (of dog or cat)
16. Maranao	sigəŋ	a peal, snore, vibration
17. Kavalan	uReŋ	the growl of a dog in anger

Note: Also Itbayaten *ahrəŋ* 'snoring; to snore', Tiruray *bugəŋ-bugəŋ* 'murmuring'. The Aklanon and Maranao forms cited here may contan a reflex of *-gəŋ 'to hum, buzz'.

*-Riŋ	to ring (dbl. *-Raŋ, *-Reŋ, *-Ruŋ)	
1. PAN	*daRiŋ	to sound
2. PWMP	*deRiŋ	to ring, buzz
3. PAN	*giRiŋ	a ringing sound
4. Sangir	bahiŋ	to snore
5. Malay	cəloreŋ	metal bar or thin plate that gives out a ringing sound when struck
6. Toba Batak	diriŋ	to ring, sound clearly
7. Malay	gəriŋ	a spherical bell
8. Lun Dayeh	kəriŋ	tiny bells tied around the ankles
9. Malay	ñariŋ	clear (of utterance); shrill or high-pitched (of the voice)
10. Arosi	ŋiri	to whine, of a dog; mew, of a cat; hum, buzz, of insects
11. Malay	pəreŋ	a ringing sound
12. Malay	reŋ	a ringing sound
13. Iban	səriŋ	a sharp, clear sound
14. Arosi	tari	to beat a gong
15. Kayan	uhiŋ	a bell

*-Ris		to drip (dbl. *-bis, *-bus.₂)
1. PMP	*tiRis	to drip, ooze through
2. Iban	bəris	shower, drizzle, spray
3. Maranao	bigis	to drip, ooze
4. Malay	diris	to water
5. Maranao	igis	to drip, dribble
6. Tombonuwo	moris-poris	to drizzle, drizzling rain
7. Tiruray	rembiris	drizzling
8. Ilokano	wáris	to sprinkle, to scatter liquids over in small drops; sow rice seeds

*-Rud		to scrape (dbl. *-rud)
1. PWMP	*aRud	to scrape off
2. PMP	*kaRud	to scrape, grate, rasp
3. PWMP	*saRud	to scrape, rub against
4. Binukid	lagʔud	to rub hard, scrape (s.t. across or against s.t.)
5. Mansaka	pigod	to scrape (as rattan)

*-Ruŋ		to roar, rumble (dbl. *-Raŋ, *-Reŋ, *-Riŋ)
1. PAN	*deRuŋ	distant rolling thunder
2. PWMP	*geRuŋ	a rumbling or low murmuring sound
3. PWMP	*heRuŋ	to roar
4. PMP	*huRuŋ	roar (of a current, crashing waves, etc.), to moan, groan
5. PAN	*keRuŋ	resounding sound
6. PAN	*qaRuŋ	sound of growling, snoring, etc.
7. PWMP	*quRuŋ	to roar, moan
8. Arosi	akuru	to thunder; thunder
9. Tolai	baruŋ	to snore, grunt
10. Roviana	buruŋu	to roar or howl, as the wind
11. Maranao	dagoʔoŋ	thunder
12. Arosi	guru	to growl, of a dog; mew, hum, buzz, hum a song or chorus
13. Soboyo	mahuŋ	to snore
14. Iban	pəroŋ	howling (as a dog)
15. Pazeh	pukakuruŋ	to rattle, clatter (the sound of a cart running)
16. Iban	rəroŋ	howling
17. Kayan	tehuŋ	the barking of dogs

The *s-roots

*-sak		ripe, cooked
1. PAN	*qesak	ripe, cooked; ready to eat
2. PAN	*Sasak	ripe, cooked
3. PMP	*tasak	ripe, cooked
4. Melanau (Mukah)	isak	to cook
5. Kelabit	laak	ripe (fruits); cooked
6. Nggela	moa-moha	cooked

*-sed		to dive, plunge into water
1. PWMP	*sesed	to plunge, dive into water
2. Kenyah	leset	to dive
3. Kenyah	meset	to dive
4. Manggarai	moset	to sink
5. Bintulu	pəsəd	to dive

*-sek$_{-1}$		to cram, crowd
1. PMP	*besek	crowded
2. PWMP	*da(ŋ)sek	to crowd, push together
3. PWMP	*desek	to press down, press together
4. PMP	*esek	crowded; to crowd together
5. PMP	*hasek	to press in, cram, crowd
6. PWMP	*isek	to cram, crowd; be compressed or congested
7. PWMP	*lusek	to cram, pack in
8. PAN	*seksek	to stuff, cram in
9. Banjarese	barasak	crammed full
10. Manggarai	cecek	to stuff, fill up
11. Kankanaey	dinsə́k	thick, close together, as plants sown thick
12. Bikol	dusók	compact, compressed, condensed, packed
13. Maranao	ləsak	to dent, press down
14. Bikol	mamsók	to be filled with milk, referring only to the teats of animals or the breasts of women
15. Kankanaey	məsə́k	laden with fruit, grain; fruitful
16. Manggarai	picek	crowded; narrow; crammed in
17. Ilokano	pusék	dense, close, crowded together, packed

(continued)

18. Tiruray	rasək	to press against, crowd
19. Javanese	rusək	crowded; to crowd, jam
20. Cebuano	tágsuk	to pack elongated things tightly in an upright position
21. Balinese	tisək	thronged, pushed close together
22. Tontemboan	wisək	to press in well (as tobacco stuffed into a pipe bowl)
23. Buginese	wusek	to stuff (as a pillow with kapok)

Note: Also PPH *saksák 'to stuff, pack tightly'.

*-sek.$_2$		to insert, stick into a soft surface (dbl. *-suk)
1. PAN	*Cesek	to insert, pierce, penetrate
2. PMP	*desek	to thrust into, pierce with
3. PMP	*hasek	to dibble, plant seeds with a dibble stick
4. PPH	*hesek.$_1$	to plant seeds by dibbling
5. PAN	*hesek.$_2$	to drive in stakes or posts
6. PMP	*isek	to insert, put in
7. PWMP	*lesek	to press into
8. PAN	*pasek	wooden nail, dowel; to drive in a wooden nail or dowel
9. PMP	*pusek	to enter, insert
10. PWMP	*tasek	insert(ed) vertically into the ground
11. PMP	*u(n)sek	to press into
12. Itbayaten	asək	putting s.t. in a hole
13. Sundanese	bələsək	to sink into the ground
14. Manggarai	cicek	to slip through a crowd
15. Buruese	kuse-k	to hammer or ram (a piling or pole into the earth); plant (a post) by digging a hole for it
16. Ilokano	lisék	to pierce the heart of an animal (in slaughter)
17. Kankanaey	ludsák	to stick in, fix in, thrust in; insert penis in vagina
18. Keley-i	luʔhək	a stake, e.g. used in holding growing beans upright
19. Manggarai	macek	to embed in
20. Bintulu	nasək	clothing
21. Kankanaey	padsák	to stick in, fix in, thrust in
22. Ngaju Dayak	pesek	pierced (esp. of ears, nose)
23. Lau	ʔoto	to pierce, insert
24. Ilokano	seksék	to penetrate to the innermost part

25. Kenyah	telasek	to join (pieces of wood), to put
26. WBM	telesek	to push s.t. into the ground
27. Melanau (Balingian)	yasek	clothing

Note: PMP *pusek was reconstructed erroneously as *pucek in Blust (1986). These two homophonous roots are semantically quite similar, but close attention to the glosses suggests that they are distinct. The inclusion of 'clothing' here is because in many AN languages one 'enters' one's clothing when putting it on.

*-sem	sour	
1. PMP	*e(n)sem	sour
2. PWMP	*isem	sour
3. PWMP	*lasem	sour
4. PWMP	*qalesem	sour
5. PAN	*qaRsem	sour
6. Bikol	tigsóm	sour; curdled (as milk)

Note: Also Amis ʔacicim 'sour'.

*-sep	to sip, suck	
1. PAN	*hisep	to suck, inhale
2. PAN	*qesep	to suck
3. PMP	*qisep	sucking, soaking up, absorbing
4. PAN	*sepsep	to suck
5. Amis	həcəp	to soak liquid up into itself
6. Subanen (Sindangan)	iksep-en	to suck
7. Puyuma	ʔasəp	to suck s.t. and swallow the juice
8. Amis	rəcəp	to soak into
9. Maranao	sasəp	to suck
10. Ayta Abellen	tolihəp	to suck through a straw
11. Cebuano	túsup	to suck the juice out of s.t.

*-seq	to wash clothes; wet	
1. PAN	*baseq	to wash clothes
2. PMP	*beseq	wet; to wash
3. PMP	*biseq	wet; to wash the anus after defecating
4. Tingalan	alisaʔ	wet
5. Javanese	asah	to wash dishes for someone
6. Sasak	bəbasaʔ	the clothes one bathes in

(continued)

7. Balinese	iŋsah	to wash rice, put through water
8. Bimanese	mbeca	wet
9. Kalagan	man-saʔsaʔ	to wash clothes
10. Sundanese	sɨsɨh	to wash clothes
11. Bulusu	usaʔ	wet

Note: Also Ma'anyan *wehuʔ* 'wet'.

*-siŋ		to spin around
1. PWMP	*gasiŋ	spinning top
2. Kayan	pasiŋ	spinning, revolving
3. DPB	pəsiŋ	dizzy
4. Malay	pusiŋ	rotate; vertigo

*-sir		hissing sound (dbl. *-sit)
1. PAN	*qesir	to hiss, sizzle
2. Malay	dəsir	the hiss of water turning into steam
3. Toba Batak	disir	to hiss
4. Nias	isi	to sough, of the wind
5. Kayan	kesih	singing of water about to boil
6. Maranao	sagisir	to hiss
7. Balinese	sirsir	to blow (breeze); to hiss
8. Makassarese	tisiʔ	to make a hissing sound with the mouth when eating hot foods, such as chili peppers
9. Ngadha	usi	to sizzle, whistle

*-sit		to hiss, sizzle (dbl. *-sir)
1. PPH	*sagitsit	to hiss, sizzle
2. PMP	*sit	the sound of sputtering, sizzling, etc.
3. PMP	*sitsit	to hiss
4. Keley-i	isit	sound made by small animal, as mouse squeak
5. Malay	ləsit	gentle rustle, such as that made by a grasshopper in grass; the buzz of angry wasps

(continued)

6. Tontemboan	lalisit	the sizzling sound of burning wood or the contents being cooked in bamboo
7. Kapampangan	salitsít	to ooze fat, sizzle

*-suD		to budge, move a bit
1. PMP	*isud	to budge, shift, move
2. Malay	aŋsur	a little at a time, in action or motion
3. Manggarai (West)	hisut	to move a little, budge
4. Minangkabau	tər-kaŋsur	edged aside
5. Bikol	ʔugsód	to be pushed or carried along by the wind or current

Note: Also PMP *icud 'to budge, shift, move aside', Malay aŋsur 'a little at a time, in action or motion'.

*-suk		to insert, penetrate, enter (dbl. *-sek₋₂)
1. PMP	*asuk	to enter
2. PMP	*buRsuk	to drive or force into
3. PMP	*dusuk	to enter, penetrate
4. PWMP	*hasuk	to enter, penetrate
5. PMP	*masuk	to enter
6. PAN	*pasuk	to enter
7. PWMP	*ra(ŋ)suk	to insert, put or drive in
8. PMP	*rusuk	to pierce, stab
9. PPH	*saluksuk	to insert, slip between
10. PAN	*suksuk	to pierce, penetrate; insert
11. PPH	*suʔuk	to enter
12. PWMP	*tisuk	to stab, stick into
13. PWMP	*tulesuk	to prick or poke with a pointed object
14. PPH	*tuRsúk	to skewer s.t.
15. PAN	*tusuk	to pierce; skewer or string together
16. PMP	*usuk	to press into, penetrate
17. Ayta Abellen	bahokhok	to insert
18. Banggai	bansuk	housepost
19. Kayan	belasuk	to push along, push into
20. Selaru	bésuk	to penetrate (of liquids)
21. Maranao	bisok	to poke, inject; injection
22. Javanese	blusuk	entering (as a mouse entering its hole)
23. Tagalog	bulúsok	sudden sinking of a foot into soft, deep mud

(continued)

24. Waray-Waray	dasók	to keep or insert s.t. haphazardly or hastily
25. Kamarian	hesu	putting on (of clothes)
26. Ida'an Begak	lasuk	to enter (of strangers or thieves)
27. Keley-i	ləhuk	to dig soil with a shovel or a wooden spade
28. Buruese	lusu-h	to cause to enter; insert
29. Kambera	nuhuku	to enter
30. Mansaka	ogsok	to stick s.t. into the ground
31. Kambera	puhuku	to enter
32. Puyuma (Tamalakaw)	Resuk	to fix a stick or post in the ground
33. Cebuano	tágsuk	to pierce; plant into but not through
34. Balinese	təlusuk	to bore a hole in the nose of a draught animal
35. Ayta Abellen	tibhok	to inject, give an injection
36. Kayan (Uma Juman)	tohuʔ	to pierce
37. Uma	-tohuʔ	to pierce

Note: Also PMP *cukcuk 'skewer'. The forms in Kambera, Cebuano, Hiligaynon and Tausug may contain a reflex of *-sek 'to insert, stick into a soft surface'.

*-suŋ		rice mortar
1. PMP	*esuŋ	rice mortar
2. PMP	*lesuŋ	rice mortar
3. PWMP	*li(n)suŋ	rice mortar
4. PAN	*Nusuŋ	rice mortar
5. Seediq (Truku)	duhuŋ	rice mortar
6. Erai	knehun	rice mortar

*-suR		satiated, full after eating
1. PAN	*besuR	satisfied from having eaten enough
2. PAN	*bisuR	satisfied from having eaten enough
3. POC	*masuR	satiated, full after eating (+ Sabah)
4. Bulusu	asug	full (stomach)
5. Yakan	esso	full (of stomach)
6. Tidung (Malinau)	masug	full (stomach)
7. Maranao	pakaosog	satisfying, of food; nutritious
8. Tidung (Kalabakang)	wasug	to be full (of stomach)

The *t-roots

*-tak.₁		mud; earth, ground (dbl. *-cak.₁, *-Caq, *-tek)
1. PCEMP	*bitak	mud
2. PWMP	*butak	mud
3. PWMP	*latak	mud
4. PSS	*litak	earth, ground
5. PWMP	*lutak	mud
6. PMP	*pitak	mud
7. PWMP	*putak	muddy
8. Toba Batak	bustak	deep mud
9. Tonbonuwo	o-gitak	a bit muddy
10. Sundanese	litak	mud, slime, mire
11. Tetun	nakutak	muddy
12. Banggai	petak	mud, muddy

*-tak.₂		sound of cracking, splitting, knocking
1. PPH	*beRták	to hit, strike, of hard objects colliding
2. PAN	*betak	to split
3. PMP	*bitak	to break, split
4. PMP	*ketak	to cluck or cackle
5. PPH	*kuták	to cackle, of a hen
6. PWMP	*latak	to hit, making a loud sound
7. PMP	*le(n)tak	to clack the tongue
8. PWMP	*letak	to split, crack
9. PPH	*litak/litaʔak	to click, clatter
10. PWMP	*pe(R)tak	to knock; pop, burst
11. PMP	*pitak	to split
12. PWMP	*qantak	to stamp, put one's foot down
13. PMP	*qetak	to slam down, ram down
14. PWMP	*Retak	to split, crack (intr.)
15. PMP	*tak	bang!
16. PAN	*taktak	to clatter, clack
17. Bontok	dəlták	to split a log without using wedges
18. DPB	dətak	to make a clacking sound
19. Bikol	gaták	split, cracked
20. Malay	gəmərətak	continuous clatter
21. Cebuano	hágtak	make a cracking or banging sound (of objects on colliding or falling)
22. Javanese	kluṭak	a clatter
23. Murut (Timugon)	kuratak	to make a clattering sound
24. Karo Batak	kurtak	to make a clicking sound with the tongue to urge a horse forward
25. Bikol	lagaták	sound of a dull smack, a slamming door

(continued)

26. Bikol	lapaták	to fall with a splat, as large drops of rain
27. BM	orotak	a plopping sound, as made by falling fruit
28. Javanese	oṭak-aṭik	to ram, batter
29. Tagalog	palaták	clacking of tongue or smacking of lips
30. Kankanaey	pátak	to begin raining; the rain comes down in thick drops
31. Gorontalo	peletaʔo	the sound of a person being struck by a flat object
32. Manggarai	rantak	to chatter (of teeth, because of cold)
33. Bikol	ruták-ruták	the cracking sound of body joints
34. Iban	satak	to knock off or out (as a knife to get the handle off)
35. Kankanaey	səgták	to knock out, collide with, strike against (as two spinning tops)
36. Toba Batak	siltak	to burst, split, crack
37. Amis	talaktak	a pounding headache; the dripping of water
38. Old Javanese	taluktak	a bamboo device, set in motion by water and making a rattling sound (used in the rice fields to scare away birds)
39. Bikol	taʔák	to split with an axe; to cleave
40. Ilokano	tarakátak	the sound of heavy rain, of water falling from the eaves, etc.
41. Tiruray	tərətak	the sound of chopping s.t.
42. Itbayaten	valtak	being cracked open, splitting, halving

*-taŋ		clanging sound (dbl. *-Cuŋ, *-teŋ₋₁, *-tiŋ)
1. PWMP	*kulintaŋ	an ensemble of gongs, the gamelan
2. PMP	*taŋ	the sound of a clang or bang
3. PPH	*taŋtáŋ	clanging sound; to clang, as in beating metals
4. Isneg	bátaŋ	an audible echo
5. Malay	dəntaŋ	to clang, e.g. of metal
6. Ngaju Dayak	garuntaŋ	scarecrow, noise-making device in the fields to frighten off pigs, deer, etc.
7. Javanese	grantaŋ	xylophone-like gamelan instrument (t for expected ṭ)
8. Malay	ləntaŋ	to clang
9. Kankanaey	litáʔaŋ	to clang, clank, ring, clink

*-taq		raw; green, of plants
1. PWMP	*bataq	young, esp. of vegetation
2. PWMP	*hataq	raw, uncooked
3. PMP	*qamataq	raw; to eat s.t. raw
4. PAN	*qataq	to eat s.t. raw
5. PAN	*qetaq	to eat s.t. raw

*-taR		flat, level
1. PWMP	*ataR	flat, level
2. PAN	*dataR	flat; flat or level land
3. PWMP	*pataR	level ground, plain
4. Bontok	datal	level; to level s.t.
5. Kankanaey	lantág	plain; flat spot, open country

*-tas		to sever, rip apart, cut through; short cut (dbl. *-tes, *-tus)
1. PMP	*bentas	to hack a passage through, blaze a trail
2. PMP	*beRtas	to tear, rip open (as cloth or stitches)
3. PMP	*bitas	to tear, rip
4. PWMP	*butas$_1$	to cut through, sever
5. PWMP	*butas$_2$	to separate people, disperse a gathering
6. PWMP	*etas	to slash, chop through
7. PWMP	*ge(n)tas	to break, break off
8. PWMP	*ketas	to cut, sever
9. PWMP	*la(n)tas	go directly, take a shortcut
10. PPH	*litas	to tear, rip (as cloth)
11. PWMP	*pi(n)tas	to cut across, cut off
12. PWMP	*qantas	to cut through
13. PWMP	*qutas	to cut through, divide by cutting
14. PMP	*ra(n)tas	to cut through, tear
15. PWMP	*retas	to split (of a seam)
16. PWMP	*ri(n)tas	to cut or tear off
17. PMP	*rutas	to sever, tear
18. PMP	*Ratas	to sever, tear
19. PWMP	*Retas	to separate gently, as in weaning
20. PWMP	*Ri(n)tas	to tear open, rip
21. PWMP	*tas	crack, snap!
22. PAN	*tastas	to sever, cut through, rip out stitches
23. PMP	*tetas	to rip or tear open
24. Tagalog	ʔagtás	an opening that is a possible path in dense forests
25. Tagalog	bagtás	passage or trail across wilderness

(continued)

26. Cebuano	báktas	to cut across (take a short cut)
27. Banggai	bantas	to harvest rice
28. Tagalog	bigtás	unstitched
29. DPB	burtas	to break, of a dyke or dam
30. Manggarai	detas	to cut something that is tied until it comes loose
31. Tiruray	fagutas	game in which contestants cut through a bundle of reeds with one
32. Kankanaey	kísat (+M)	the sound of tearing up cloth
33. Dumagat (Casiguran)	lagtás	to chop in two (a vine, rope or string)
34. Sambal (Botolan)	ləntah	to come unsewed, rip or cut into
35. Cebuano	lúgtas	to break a rope, thread, string, etc. by pulling on it with force
36. Binukid	lutas	to wean a child
37. Chamorro	motas	defect in sewing, weak stitching
38. Rejang	nitas	to take a shortcut
39. Uma	pataʔ	to cut through, sever
40. Kankanaey	piŋsát	to tear, rend, lacerate, rip up
41. Abaknon	putas	to cut with one stroke
42. Bikol	rugtás	to pull or tear s.t. apart; to tear down
43. Bikol	rutás	to end or terminate (a talk, sermon, or sermon)

Note: Cebuano *báktas* may reflect earlier *bagtás*, with voice assimilation.

*-taw$_1$		to float
1. PWMP	*la(n)taw	to float
2. PWMP	*le(n)taw	to float
3. PWMP	*lutaw	to float
4. PWMP	*paletaw	to float
5. PPH	*pataw	buoy
6. PPH	*Ra(n)taw	to float
7. PPH	*tagal(e)táw	buoy; to float
8. Dumagat (Casiguran)	baltáw	for a fish or crocodile to swim along on the surface of the water
9. Cebuano	hátaw	to appear for a moment on a surface
10. Ayta Abellen	kotaw	to float
11. Tagbanwa (Central)	liminutaw	to float
12. Bare'e	nanto	to float, to raft
13. Bikol	runtáw	to appear on the surface of the water (fish)
14. Maranao	taltaw	buoyant
15. Cebuano	utaw-útaw	to float

*-taw.₂		lost, astray (dbl. *-gaw, *-ŋaw.₁)
1. PWMP	*tawtaw	lost, astray
2. Cebuano	kútaw	get disturbed, confused
3. Kankanaey	na-lítaw	lost; stray; forlorn
4. Kankanaey	na-motáw	lost, stray; forlorn
5. Keley-i	tawitaw	for someone to become confused

*-tay.₁		to hang
1. PPH	*bítay	to hang
2. PWMP	*quntay	to dangle, hang down loosely
3. Maranao	bəntay	to hang
4. Cebuano	hutáy	line hung for drying; to string a line for drying
5. Tausug	juntay	a few strands of hair hanging loosely (from a woman's bun), hair hanging down in strands
6. Bikol	kulambitáy	to hang by the hands (as from a branch)
7. Ibaloy	oktay	to hang out, as a handkerchief hanging out of someone's pocket

*-tay.₂		suspension bridge
1. PWMP	*kitay	suspension bridge
2. PWMP	*lantay	bridge, floor
3. PCMP	*letay	bridge
4. PMP	*taytay	single log bridge
5. PMP	*titay	suspension bridge
6. Itawis	balátay	bamboo bridge
7. Nias	ete	small bridge
8. Tetun	lalete	a light bridge
9. Keley-i	laŋtay	bridge
10. Kambera	líndi	bridge
11. Mentawai	nitay	bridge
12. Bontok	ʔalátəy	a bridge, usually a single log across a small stream
13. Molima	ʔitete	bridge
14. Ilokano	raŋtáy	bridge

Note: There is no doubt that *-tay.₁ 'to hang', and *-tay.₂ 'suspension bridge' are ultimately the same root. However, there is no overlap in either form or meaning for the forms given here, so each is cited separately.

*-teb		to prune; graze
1. PWMP	*e(n)teb	to cut off, prune
2. PMP	*keteb	to bite
3. PAN	*tebteb	to cut
4. Maranao	gatəb	to gnaw, chew
5. Maranao	kitəb	to grind, sharpen, gnash the teeth
6. Bontok	ʔaŋtəb	to graze, of cattle
7. Maranao	rətəb	to gnaw, nibble
8. Maranao	tatəb	to graze, browse, feed, eat, bite, gnaw

*-tek.₁		a clicking or light knocking sound
1. PWMP	*ketek	the sound of dull clucking
2. PWMP	*letek	clattering sound
3. PWMP	*patek	clicking sound
4. PMP	*qentek	to stamp on
5. PWMP	*sale(n)tek	clicking sound
6. PMP	*tek	thump!
7. PAN	*tektek.₁	chopping to pieces, cutting up, as meat or vegetables
8. PAN	*tektek.₂	gecko; chirp of gecko
9. Ibatan	bitek	beating of the heart
10. Ngaju Dayak	garetek	knocking, sound of knocking
11. Sundanese	gələtək	the sound of commotion
12. Tboli	getek	an abrupt sound, as ticking of a clock
13. Ilokano	kiték	to tick (as clocks do)
14. Javanese	kluṭak	clatter; make continuous soft sounds (as a mouse gnawing in the kitchen)
15. Melanau (Mukah)	latək	to hammer (a nail, etc.)
16. Ilokano	lítek	noise made by finger joints; clicking noise (as when taking pictures)
17. Bontok	palpaltak	k.o. frog (onom.?)
18. Javanese	pərkətək	cracking or crackling sound
19. Dumagat (Casiguran)	pitək-pitək	to click, to throb, to tick, to make a sound like the ticking of a clock or the throbbing of one's pulse
20. Puyuma	ʔatək	to cut down (usually a tree)
21. Ibaloy	shitek	to strike a tethered cow or carabao in the back of the neck with an axe
22. Maranao	təntək	to cluck, as hen calling chicks; make a sound with the tongue
23. Tiruray	tərəktək	a house lizard

*-tek₂		mud, muddy (dbl. *-cak₁, *-Caq, *-tak)
1. PMP	*litek	muddy; sticky, as mud
2. PPH	*lutek	mud
3. PPH	*pitek	mud
4. Kayan	atek	muddy or discolored—of water, etc.
5. Rembong	gitek	mud, muddy
6. Old Javanese	itək	to live (play, wallow) in the mud
7. Old Javanese	latək	mud, bog

Note: Also Cebuano *hútik* 'to be muddy; for an area to turn to mud', PPH *putik 'mud'.

*-teŋ₁		to hum, drone (dbl. *-Cuŋ, *-taŋ, *-tiŋ)
1. PMP	*teŋ	to buzz, hum
2. PAN	*teŋteŋ	to drone, droning tone
3. Isneg	battáŋ	to beat (said of the heart)
4. DPB	dətəŋ	to make a humming sound
5. Ifugaw	getéŋ	rattling of metals that collide against each other
6. DPB	kərtəŋ	the hum made by a wasp when it comes to attack
7. Ilokano	kuteŋtéŋ	the sound of the guitar

Note: Also Ilokano *weŋweŋ* 'a buzzing sound produced in the ear', *yeŋyeŋ* 'bothered, confused, bewildered, flurried (with noise)'.

*-teŋ₂		to stare, look fixedly (dbl. *-Neŋ, *-Nuŋ)
1. PMP	*e(n)teŋ	to stare, look fixedly at
2. PWMP	*teŋteŋ	to stare, look fixed at s.t.
3. Ida'an Begak	atoŋ	to look around
4. Tombonuwo	moŋo-bontoŋ	to stare
5. Maranao	patəŋ	to stare
6. Bikol	pítoŋ	to stare at s.t. with awe or rapture

Note: Also Lun Dayeh *tutoŋ* 'a look or stare'.

*-teŋ₃		to stretch; taut (dbl. *-deŋ)
1. PMP	*benteŋ	extended, stretched taut
2. PWMP	*keteŋ	to straighten out, of a limb, etc.
3. Tagalog	bagtíŋ	taut
4. Maranao	bintəŋ	to stretch
5. Malay	gətaŋ	astretch; taut

(continued)

6. Tagalog	ʔigtín	taut
7. Mapun	intoŋ	straight
8. Maranao	korintəŋ	to stretch
9. Isneg	pattáŋ	to tighten, to stretch
10. Javanese	pəṭəṇṭəŋ	tensed with effort or anger
11. Maranao	tagəntəŋ	to stretch
12. Makassarese	tantaŋ	stretched taut; steadfast

*-teq	sap, gummy secretion	
1. PAN	*diteq	sticky substance
2. PWMP	*gateq	sap, resin
3. PWMP	*geteq	sap, gummy secretion
4. PWMP	*giteq	tree sap
5. PMP	*liteq	sap of a tree or plant
6. PWMP	*muteq	gummy secretion of the eyes
7. PAN	*Niteq	sap of a tree or plant
8. PAN	*puteq	gummy secretion
9. Tagalog	dagtáʔ	sap, resin
10. Tagalog	gitatáʔ	sticky with moist dirt
11. Maranao	kamontaʔ	sticky sap
12. Sasak	tataʔ	to tap latex for birdlime

*-ter	to shiver, tremble (dbl. *-ger, *-tiR)	
1. PMP	*eter	to shake, vibrate, tremble
2. PWMP	*keter	to shiver, quiver
3. PAN	*terter	to shiver, tremble
4. Manggarai	denter	to quiver, tremble (of a wall)
5. Sundanese	gətər	to quake, tremble
6. Singhi	kimatur	to tremble

Note: Also Ilokano *ag-tayegtég* 'to shake, tremble'.

*-teR	stiff, rigid, as a corpse (dbl. *keR)	
1. PPH	*ke(n)teR	stiff, rigid
2. Pangasinan	kígtəl	become stiff, set
3. Manggarai	lenter	stiff, rigid (of a corpse)
4. Manggarai	nter	stiff, taut

Note: Also Ida'an Begak *kətur* 'stiff (of hair), rigor mortis'.

*-tes	to tear, rip (dbl. *-tas, *-tus)	
1. PWMP	*butes	a broken piece of rope or string
2. PMP	*ge(n)tes	to snap, break
3. PMP	*qetes	to chop, hack, cut off
4. PWMP	*re(n)tes	to separate, remove
5. PMP	*tes	to tear
6. PWMP	*testes	to rip, tear, shred
7. Old Javanese	gitəs	to break off, pick
8. Maranao	kətəs	cutting
9. Sundanese	pitəs	to snap or break to pieces
10. Sasak	putəs	to break (of a rope)

*-tik.₁	gentle curve	
1. PPH	*lantik	graceful curve (of body)
2. PMP	*lentik	graceful curve
3. PWMP	*palantik	curved
4. Sundanese	bəntik	an unnatural curve of the body, as of the elbow bending inward
5. Malay	cantek	pretty; fair to look upon
6. Sundanese	cəntik	bent, curled back, as the toes of slippers, or the eyelashes

*-tik.₂	small, little, few (dbl. *-dik)	
1. PWMP	*etik	little, few, small amount
2. PAN	*qitik	small, little, few
3. Kankanaey	atík	a little, a few
4. Isneg	bittíʔ	little, small, few
5. Sundanese	jəntik	little finger, pinkie
6. Minangkabau	kéték	small
7. Ibaloy	otik	to become small, few
8. Javanese	siṭik	a few, a little
9. Rejang	titiʔ	small

Note: Also Sarangani Manobo *deitek* 'small, tiny', Dibabawon Manobo *ma-ʔíntek* 'small (object)', Binukid *yentek* 'tiny, very small; fine, not course'.

*-tik₃		ticking sound	
1.	PWMP	*detik	to throb, tick
2.	PMP	*gi(n)tik	to tap or beat lightly on
3.	PWMP	*hag(e)tik	the sound of ticking
4.	PMP	*katik	high-pitched sound of striking against s.t. hard but not resonant
5.	PMP	*ketik	to make a noise
6.	PMP	*kitik	to tick; to make a ticking or light knocking sound
7.	PWMP	*letik	the sound of snapping or cracking
8.	PMP	*li(n)tik	to snap off; snapping or clicking sound
9.	PWMP	*pa(n)tik	the sound of ticking, flipping, or light tapping
10.	PAN	*peCik	clicking sound
11.	PMP	*pi(n)tik	to throb, beat
12.	PWMP	*santik	stone for striking fire
13.	PWMP	*simetik	to tick, click
14.	PPH	*sintik	to knock two things together with a clicking sound
15.	PPH	*tariktik	woodpecker
16.	PWMP	*taRiktik	the sound of ticking or raindrops falling
17.	PAN	*tik	the sound of tapping, flicking the finger against s.t.
18.	PAN	*tiktik	to tap, strike lightly on a hard surface
19.	Malay (Sumatra)	bəlatek	woodpecker
20.	Tetun	betik	to click the fingers, fillip
21.	Ilokano	bitík	to palpitate, throb, flutter, pulsate violently
22.	Ilokano	bulintík	to play at marbles
23.	Fijian	calidi	to burst, explode, detonate
24.	Balinese	əṭik	to knock, tap, rap
25.	Old Javanese	gaṭik	to strike, clash
26.	Balinese	gəṭik	to make a knocking sound (t for expected ṭ)
27.	Ifugaw	hultík	marbles
28.	Yamdena	itik	the beating of heart or pulse
29.	DPB	kəritik	the sound of sand ground between the teeth, oil sizzling in a frying pan
30.	Malay	kərtek	a continuous ticking, the sound of a clock
31.	Javanese	kluṭik	a metallic plink
32.	Dusun (Kadazan)	koitik	to rap, to knock
33.	Palawano	kuritik	small stick used to hit the rim of the gong with a fast tempo
34.	Ilokano	labtík	pulsation, beat, throb (heart, arteries)

(continued)

35. Bikol	lagatík	a ticking sound
36. Tagalog	lagitík	to creak, creaking (<*-tek₁?)
37. Bikol	lapatík	sound of tapping (as a finger nail on a window pane, or a pebble thrown at glass; to tap, patter
38. Tiruray	latik	to snap the fingers
39. Fijian	lavidi	a slight ticking, cracking of joints
40. Bikol	lubtík	to snap (rubber band, bowstring)
41. Gorontalo	poloti?o	pulse; blow
42. Ifugaw	putík	the sound produced by a boulder that bursts
43. Javanese	rəkitik	a light crackling sound, e.g. of paper burning (*t* for expected *ṭ*)
44. Iban	ratik	(of rain) pattering, rattling
45. Itbayaten	ritich	noise produced when pouring water on fire, or popping corn
46. Puyuma	saltik	to strike a match
47. Ibaloy	seltik	to snap one's fingers
48. Iban	sətik	to click
49. Sundanese	talektek	to tap or drum on s.t. aimlessly
50. Ilokano	tarakítik	the sound of drizzling rain, etc.
51. Lamaholot	tətik	to drip
52. Cebuano	witík	to strike something with a flick

*-til₁		to pinch off a bit
1. PCMP	*bitil	to pinch (as to pick up a small amount of some substance)
2. PAN	*getil	to pinch
3. PAN	*ketil	to pinch off
4. Dumagat (Casiguran)	kítil	to pinch off the tops of leafy vegetables for cooking

*-til₂		a small protruding part
1. PWMP	*butil	grain, kernel
2. PWMP	*itil	clitoris
3. PWMP	*kutil	wart
4. Itbayaten	atil	clitoris
5. Sundanese	bəntil	pimples
6. Malay	bintil	pustule; pimple; heat-spot; mosquito bite

(continued)

7. Kapampangan	bunítil	one grain of rice
8. Sundanese	cəntil	to protrude, as a wart on s.o.'s cheek; also, the last remaining tooth in s.o.'s mouth
9. Malay	pəntil	sprout, rudimentary fruit
10. Javanese	səntil	uvula; a small soft protuberance

Note: Also Toba Batak *binsil* 'clitoris'.

*-tiŋ		a clear ringing sound (dbl. *-Cuŋ, *-taŋ, *-teŋ)
1. PPH	*batiŋtiŋ	ringing sound, as of a bell
2. PMP	*tiŋ	a clear ringing sound
3. PMP	*tiŋtiŋ	a tinkling sound
4. Hiligaynon	bágtiŋ	to ring a bell, to toll
5. Tiruray	bantiŋ	a bell
6. Subanen/Subanun	bastiŋ	a bell
7. Maranao	bətiŋ	a bell; to ring a bell
8. Malay	dəntiŋ	to chink or clink – of coin or thin metal
9. Aklanon	eántiŋ	high-pitched (voice)
10. Bikol	karagtíŋ	shattering sound, such as that made by a fallen plate, coin, spoon; clinking sound
11. DPB	kərtiŋ	ringing sound, as of a coin that is dropped
12. Malay	kəruntiŋ	cattle-bell (of wood)
13. Yakan	lagtiŋ	piercing, shrill, high pitched (of sound, tone, voice)
14. Malay	lətiŋ	a tinkle such as the note of a mandoline or guitar
15. Ilokano	litíŋ	shrill sound
16. Manggarai	rentéŋ	tuneful, melodious (of the voice of a person or animal)
17. Maranao	talintiŋ	shrill

*-tip		tongs; to pinch together
1. PMP	*qatip	to pinch, squeeze, press together
2. DPB	gatip	rice harvesting knife (used in palm, with thumb pressing against blade)
3. Hawu	gəti	to pinch with the nails
4. Buli	latif	to hold with tongs
5. Singhi	sotip	pincers

*-tiR		to tremble, quiver (dbl. *-ger, *-ter)
1. PWMP	*tiRtiR	to shake, shiver, quake, tremble
2. Sika	betir	to flutter (as the wings of a caged bird)
3. Tarakan	talintig	to quake, shiver
4. Manggarai	tetir	to flutter, shake, shiver

*-tuk$_1$		beak
1. PWMP	*pa(n)tuk	beak of a bird
2. PAN	*tuktuk	beak of a bird; to peck
3. Rejang	catuʔ	a beak
4. Malay (Sumatra)	cetok	beak of a bird
5. Malay (Jakarta)	cotok	beak of a bird
6. Tae'	titok	to peck with the bill; to bite, of a snake
7. Rejang	tuʔ	beak

*-tuk$_2$		to bend, curve (dbl. *-kuC, *-kuk, *kug, *-luk/luquk)
1. PMP	*be(n)tuk	curve
2. PWMP	*hutuk	to bend, flex
3. PAN	*latuk	curved, bowed
4. PWMP	*lentuk	to bend, flex
5. PPH	*lituk	to bend, curve
6. PMP	*tutuk	to bow, lower the head
7. Maranao	balitok	to bend
8. Kapampangan	bayútuk	to bend, sag
9. Tetun	bituk	to make an edge or border; to fold, bend or double strips of palm leaf when weaving so that it is neat and tidy
10. Mansaka	botok	to bend down (as a tree limb)
11. Fijian	kodu	to bend a thing; arch, like the back of a cat
12. Tiruray	lontuk	concavely curved
13. Hawu	ŋetu	to bend, bow, curtsy
14. Dumagat (Casiguran)	pitók	curved, bent
15. Fijian	rodu	bent, curved
16. Manggarai	wituk	swaying the body, posturing

*-tuk.₃		to nod	
1. PWMP		*kantuk	to nod the head in drowsiness
2. PWMP		*qantuk	to nod the head in drowsiness
3. Pangasinan		toʔə́k	to bow the head
4. Tagalog		tukatók	nodding the head because of sleepiness
5. Banggai		tutuk	to nod the head
6. Banjarese		untuk	to nod the head

Note: Possibly identical to *-tuk-₁ 'to bend, curve'.

*-tuk.₄		top, summit, crown	
1. PWMP		*pa(n)tuk	apex, peak
2. PWMP		*puntuk	top, tip, apex
3. PPH		*taluktuk	hummock, small rise of land
4. PMP		*tuktuk	top, summit, crown
5. PPH		*u(n)tuk	on top, above
6. Dumagat (Casiguran)		búntok	head; top of the head
7. Pazeh		ituk	on top; the top of the head
8. Amis		lotok	mountain
9. Ida'an Begak		pəmuntok	the top of a hill

Note: Also Karo Batak *pucuk* 'the top, upper end of a plant'.

*-tul		to bounce	
1. Malay		antul	to bounce, as a ball; specifically to bounce up, in contrast to *məŋ-ambul* 'to bounce back'
2. WBM		datul	of an object, to bounce
3. Malay		pantul	to bounce back, as a ball
4. Bikol		ʔuntól	to bounce; to rebound

*-tuq		to fall (dbl. *-buq.₂)	
1. PWMP		*zatuq	to fall
2. Bintulu		gatuʔ	to fall
3. Fijian		lutu	to fall from a height
4. Ida'an Begak		ratuʔ	to fall
5. Malay		rəntoh	toppling down, crashing down
6. Malay		runtoh	toppling down, crashing down
7. Berawan (Long Terawan)		sito	to fall, of fruit
8. Kelabit		tutuʔ	to fall from a height

*-tus		to snap under tension, as a string or rope dbl. *-tas, *-tes)
1. PMP	*betus	to burst open, as an overfilled sack
2. PMP	*butus	to snap, break, as a rope under tension
3. PWMP	*getus	to snap, break, as a string
4. PMP	*ku(n)tus	to break under tension
5. PWMP	*putus	to snap, break off
6. PMP	*utus	to break under tension, as a rope
7. Tontemboan	entos	to separate, of married persons
8. Cebuano	hágtus	for a line or rope to break under tension
9. Tiruray	kətus	to break a string or rope due to excess tension
10. Bikol	lugtós	to break or snap under pressure (as a string holding a kite or bird, a fishing line with a fish on the hook (< *tes?)
11. Samoan	motu(-sia)	(of long things such as a rope or stick), to break, snap
12. Balinese	tustus	to break off, pluck
13. Itbayaten	vi?tos	being cut apart by applying force or by pulling (of rope, etc.)

*-tut		flatulence; stench
1. PWMP	*be(n)tut	fart; stench
2. PPH	*buŋ(e)tut	stench, bad odor
3. PWMP	*ke(n)tut	flatulence
4. PAN	*qetut	fart, flatulence
5. PMP	*qutut	flatulence; to fart
6. Yami	atot	fart
7. Ayta Abellen	bantot	to smell bad (as stagnant water)
8. Keley-i	bəntut	bad water, e.g. smelly, stale
9. Keley-i	ʔəbtut	to have a very bad, rotten-type odor; stinking odor
10. Sundanese	hitut	fart
11. Hanunóo	laŋtút	the repulsive odor of long-standing water and the like
12. Ilokano	luŋtót	rotten, putrified, decomposed, spoiled, putrid
13. Bontok	ʔoptót	to have the smell or taste of old rice that has begun to go mouldy
14. Maranao	tagtot	to make wind due to stomach gas
15. Kankanaey	ubtút	to stink, smell badly

The *w-roots

*-waŋ		wide open space (dbl. *-baŋ₋₂)
1. PMP	*awaŋ	atmosphere, space between earth and sky
2. PWMP	*banawaŋ	open space
3. PWMP	*bawaŋ	an open expanse of land or water
4. PWMP	*gawaŋ	a gate, opening
5. PPH	*giwaŋ	gap, open space, breach
6. PWMP	*kawaŋ	to be apart, separated
7. PWMP	*lawaŋ₋₁	door, gate passageway
8. PWMP	*lawaŋ₋₂	wide open, spacious
9. PWMP	*liwaŋ	open space
10. PWMP	*rawaŋ	hollow, cavity
11. PWMP	*Rawaŋ	a gap; full of gaps
12. PMP	*sawaŋ	wide open spaces
13. PMP	*siwaŋ	gap, open space, groove
14. PWMP	*tawaŋ	open space, empty of obstruction
15. PPH	*tiwaŋwáŋ	opening, gap
16. PPH	*waŋáwaŋ	wide open space
17. PPH	*waŋwáŋ	gaping expanse
18. Motu	gadava	the space between the legs, toes and fingers
19. Yami	iwaŋ	to open, open a door
20. Kayan	jawaŋ	large-sized mesh of net
21. Malay	kərawaŋ	openwork, a' jour design
22. Lun Dayeh	kinawaŋ	a valley, a space or opening in the forest
23. Keley-i	kiwaŋ	to move to make way for someone
24. Itbayaten	lakwaŋ	air space, interval
25. Manggarai	lewaŋ	the space between the teeth; doorway
26. BM	ŋawaŋ	atmosphere, space
27. Amis	ʔadawaŋ	a gateway
28. Bikol	ruwáŋ	the open space on a boat (between seats or partitions, in compartments, etc.)
29. Ida'an Begak	sərawaŋ	to cut away plants at the border of a rice field to see the neighbors
30. Sika	waŋ	the opening of a door
31. Yami	zawaŋ	sea channel

*-waq		spider
1. PAN	*lawaq	spider, spiderweb
2. Gana	aŋkalawaʔ	spider

(continued)

3. Kadazan Membakut	aŋkukulawa?	spider
4. Itbayaten	axaawa	spider
5. Kapampangan	babagwá?	spider
6. Kanowit	bəkawa?	spider
7. Subanon	boliŋkawa?	spider
8. Lotud	iŋkaralawa?	spider
9. Dumpas	kakawa?	spider
10. Bontok	kawa	spider
11. Kenyah (Long Atun)	kəlawa?	spider
12. Melanau (Matu)	kətawa?	spider
13. Molbog	kololawa?	spider
14. Mota	marawa	spider
15. Bisaya (Sabah)	sigagalawa?	spider
16. Tarakan	taŋkawara	spider
17. Kayan (Uma Juman)	təlawa?	spider
18. Sabah Bisaya	tiŋalawa?	spider
19. Tidung (Sumbol)	tiŋkawa?	spider

*-wit	a hook; hook shaped (dbl. *-bit._1)	
1. PWMP	*banuit	fishhook
2. PPH	*bunuwit	fishhook
3. PWMP	*kalawit	a hook
4. PAN	*kawit	a hook; to hook
5. PMP	*lawit	a hook
6. PWMP	*ruit	barbed; sharp
7. PWMP	*sawit	a barb
8. Ngaju Dayak	awit	curved, hooked
9. Banjarese	bait	fishhook
10. Isneg	balawít	an arrow with an iron head; it has three barbs on one side, none on the other
11. Isneg	bilawít	an arrow and a spear whose iron head has two barbs, one on either side, but asymmetrical, one of them being cut at a higher level than the other
12. Mansaka	biŋwit	to catch with hook and line
13. Ifugaw	buŋwit	to fish with small hook and line
14. Lamaholot	diwit	to hang s.t. up
15. Tagalog	dukwít	taken out or off by a hooked tool
16. Malay	gait	hook
17. Malay	(h)uit	a barb
18. Kelabit	keluit	a fishhook
19. Hawaiian	kiwi	curved object, such as a sickle

(continued)

20.	Amis	korawit	to grasp or pull in with a hook
21.	Amis	ŋawit	to pull the trigger of a gun
22.	Toba Batak	rait	a hook
23.	Kankanaey	sakuʔít	to hang upon, hook, catch in
24.	Iban	səŋkawit	to cling to, hang
25.	Kankanaey	sokláwit	to hang (blanket, etc.) on something more or less pointed

Note: Due to the predictability of a transitional glide, -wit following *u is written -it.

The *y-roots

*-yuŋ		to shake, sway, stagger
1. PWMP	*ayuŋ	to swing, sway, rock
2. PWMP	*guyuŋ	to shake, sway
3. PMP	*huyuŋ	to shake, sway, as the ground in an earthquake
4. PWMP	*kuyuŋ	to shake, sway, stagger
5. PWMP	*luyuŋ	to sway, totter
6. PWMP	*ruyuŋ	to reel, stagger
7. PWMP	*Ruyuŋ	to shake, sway, stagger
8. Chamorro	yeŋyoŋ	to shake, vibrate

Note: Also PWMP *ayun 'to swing, oscillate', PWMP *uyug 'to shake, sway, stagger'.

*-yut		bag, carrying pouch
1. PWMP	*uyut	carrying pouch or small basket
2. WBM	beruyut	a sack or bag woven of sedge or pandanus
3. WBM	kemuyut	a cloth or woven sedge bag used to carry ingredients for the betel chew
4. Binukid	puyut	to carry s.t. in a piece of cloth, paper, etc. by pulling up the edges

The *z-roots

*-zak		step, tread, trample (dbl. *-zeg, *-zek)
1. PWMP	*anzak	to stamp the feet
2. PMP	*bezak	to step, tread, stamp on
3. PWMP	*enzak	to step, tread; stamp the foot
4. PMP	*inzak	to step on, tread
5. PCEMP	*kezak	to step on, tread on
6. PMP	*ki(n)zak	to tread, step on
7. PWMP	*lanzak	to step, tread on
8. PPH	*padák	to trample
9. PWMP	*pizak	to step, tread on
10. PMP	*re(n)zak	to step on, place one's foot on
11. PWMP	*tanzak	to stamp the feet; step, leap
12. PWMP	*ti(n)zak	to step, tread
13. PWMP	*tuzak	to step on
14. Pazeh	mu-dakedak	to thresh with feet (as millet)
15. DPB	ərjak	to step on
16. Malay	jəjak	to trample on
17. Malay	jijak	footfall; step, tread
18. Bare'e	kaja	to step with the flat of the foot and a sideways movement of the leg
19. Iban	lunjak	to trample, tread or stamp on repeatedly
20. Tagalog	paldák	hardened or flattened by repeated treading or trampling
21. Aklanon	púsdak	to stomp, bang one's feet
22. Bare'e	yanda	a tread, step
23. Agta (Eastern)	yudak	step on, stamp on, stomp on

Note: Also PWMP *qundak 'jouncing movement; stamp the feet'.

*-zam		to borrow
1. PWMP	*hinzam	borrowing, lending
2. PWMP	*huzam	borrowing
3. PAN	*Sezam	to borrow, lend
4. Muna	ada	to borrow

*-zaŋ		long, tall
1. PWMP	*kazaŋ	long; stilts
2. Sasak	enjaŋ	stilts
3. Toba Batak	ganjaŋ	long, tall

(continued)

4. Bahasa Indonesia	lanjaŋ	long and thin
5. Malay	panjaŋ	length in space or time
6. Ngaju Dayak	tanjaŋ	stature, figure (said of people)
7. Lampung	tijaŋ	long
8. Minangkabau	tonjaŋ	long, lanky

*-zeg		to stand erect (dbl. *-zeR)
1. PWMP	*ta(n)zeg	to stand erect
2. PMP	*tinzeg	to stand erect
3. PMP	*uzeg	to make erect
4. PWMP	*zegzeg	to stand erect
5. Balinese	ajəg	erect, stand upright
6. Manggarai	cedek	upright
7. Manggarai	codek	to stand or sit upright
8. Ngaju Dayak	gagedek	full, of sails
9. Mansaka	indəg	to stand
10. Singhi	kajog	to place upright
11. Singhi	mojug	to stand up
12. Old Javanese	pajəg	a "stand", arrangement, set (of s.t. standing upright)
13. Maranao	pamayandəg	to stand upright
14. Manggarai	pendek	straight, erect (of a post)
15. Mandaya	sakindəg	to stand
16. Kankanaey	takdəg	to stand, stand up
17. Manggarai	tendek	to stand upright (as bamboo, spikes)
18. BM	torindog	erect
19. Balinese	tujəg	to be erect, upright, vertical
20. Old Javanese	tulajəg	to stand erect

Note: Also PMP *tezek 'to stand erect; upright (posts)'.

*-zek		step, tread, trample (dbl. *-zak)
1. PAN	*zekzek	to tread, trample on
2. Ilokano	baddék	footprint, step; to step on s.t.
3. Balinese	ənjək	to tread on, stand on, stamp on
4. Uma	pedeʔ	to stamp on s.t.
5. Bintulu	tunjək	to step on

Note: Also Bikol *parág-padág* 'to stamp the feet repeatedly (as when a child throws a tantrum)', *purág-pudág* 'a galloping sound'.

*-zel	dull, blunt	
1. PWMP	*ŋazel	dull, blunt
2. Kayan (Uma Juman)	kasəl	dull, blunt
3. Ilokano	ŋudél	dullness
4. Sa'ban	padəl	dull, blunt
5. Melanau (Matu)	tajəl	dull, blunt

*-zem	to close the eyes	
1. PAN	*kezem	to close the eyes
2. PWMP	*kizem	to close the eyes
3. PWMP	*pezem	to close the eyes
4. Melanau (Mukah)	pajəm	closed, as the eyes; extinguished, of a fire
5. Sasak	pidəm	to close the eyes; fall asleep

Note: Also Kenyah *pidem* 'to close the eyes'.

*-zep	to blink; flicker, of light	
1. PWMP	*izep	to wink, blink
2. PWMP	*kezep	to blink, wink, momentarily close the eyes
3. PWMP	*ki(n)zep	to blink, wink
4. Kayan	jep	a blink
	jep-jep	to blink the eyes; to twinkle (of a light)
5. Kayan	pejep	flickering, blinking, of a light

*-zeR	to stand erect (dbl. *-zeg)	
1. Karo Batak	cindər	to stand, stand erect
2. Pazeh	kizex	to step up on (a rock, table, etc.)
3. Sika	sader	to stand erect
4. Sasak	tajər	stiffly erect

*-zul		to thrust out, protrude (dbl. *-zur)
1. Malay	benjol	big inflamed bump on forehead
2. Malay	rojol	to jut out (of a snake's head emerging from a hole, a man's head thrust out of a window, etc.)
3. Maranao	tədol	to protrude, project, come out
4. Malay	tonjol	bulge; protuberance; hump
5. Bikol	ʔutdól	to protrude or stick out through

*-zur		to thrust out, protrude (dbl. *-zul)
1. PMP	*qunzur	to thrust out, extend forward
2. Malay	anjur	stretching out, advanced beyond
3. Iban	bəjur	stretched out
4. Kelabit	bujur	distended, of the belly
5. Tagalog	duldól	shove or thrust s.t. with force into another, as pen into ink bottle, food into mouth, etc.
6. Melanau (Mukah)	ijuh	to stretch out the legs (as after sitting cross-legged for awhile)
7. Madurese	lanjur	exceptionally tall, 'sticking out'
8. Iban	tajur	spit of land; uncleared land extending into clearing
9. Kayan	tejoh	stretched out straight (as the legs when sitting or sleeping)
10. Bintulu	təpəju	outstretched (of legs when sitting)
11. Tboli	tidol	to stretch out, lie down full length; to have long legs
12. Manggarai	wendor	extend the legs while sitting

*-zut$_{-1}$		to pinch, pick up with thumb and forefinger
1. PWMP	*kezut	to pinch, as with the nails
2. PWMP	*kuzut	to pinch
3. PPH	*pídut	to pinch, hold between the fingers
4. PWMP	*pu(n)zut	to pick up with fingertips
5. PWMP	*sazut	to pick up
6. Tagalog	dalirot	poking or stirring up with the finger
7. Ifugaw (Batad)	hapúdut	for s.o. to pick or pull up s.t. with the fingers and thumb (as dirt or rubbish)
8. Isneg	kaldut	to be picked, said of leaves
9. Kapampangan	kandut-	to pinch with a twisting motion

(continued)

10. Pangasinan	karot	to pinch
11. Keley-i	putdut	to pick up s.t. using thumb and forefinger

Note: Also Maranao *pindit* 'to hold with or between the fingers'.

*-zut.$_2$		to uproot, pluck, pull out (dbl. *-buC, *-guC, *-NuC)
1. PPH	*badut	to uproot, pull out
2. PMP	*bezut	to uproot, pluck, pull out
3. PPH	*ladut	to pluck, pull out, uproot
4. PMP	*Razut	to pluck, pull out
5. PMP	*zutzut	to pluck, pull out
6. Pangasinan	ákdot	to remove, pluck out
7. Maranao	andot	to pull out
8. Wolio	bindu	to pull out, extract (tooth, sword, weeds, cork)
9. Maranao	dayodot	to pull out, evacuate
10. Banggai	idut	to pull off clothes
11. Tboli	kedut	to pull out, as feathers, weeds

Note: Also PMP *dutdut 'to pluck, pull out'.

6 Appendices

There are two appendices, the first a display of roots both by word class, and by semantic class/referent type, and the second a short list of root candidtates which currently are known to be supported by three EIMs, but for which a fourth has not yet been efound, despite a concerted effort to incorporate them into the dictionary.

Appendix 1

Breakdown of roots by word class (Display 1), and freqency of association with a semantic class/referent type (Display 2).

Display 1: Word classes in which roots have been found.

(A) Verbs		
to untie, unravel: 2	to skin, peel off: 3	to split: 4
to accompany: 1	to tie, wind around: 3	to plug, block, dam: 5
to twist, twine, braid: 4	to hook, clasp: 2	to whip: 1
to pull out, uproot: 4	to grow: 1	to fall: 2
to stir, mix: 4	to strew, sow: 2	to end, finish: 2
to leak, spill out: 1	to hatch: 1	to splash, squeeze out: 3
to flick, spring up: 1	to shine: 7	to lean against: 1
to think, brood: 1	to shelter, shade: 2	to grasp, grip, seize: 3
to shiver, tremble: 3	to scratch, scrape: 4	to bite: 2
to open: 2	to prop, support: 3	to spread apart, as legs: 1
to grope, feel in the dark: 1	to begin: 1	to rise, climb: 1
to bend, curve: 3	to lie prone: 2	to cough: 1
to shrink, shrivel: 1	to cover: 2	to wrap around, as snake: 1
to join lengthwise: 1	to surround: 1	to close: 1
to shine, glitter: 7	to open the eyes: 1	to sink, submerge: 7
to swallow: 1	to turn around, revolve: 1	to insert, penetrate: 3
to return, restore: 1	to flow: 2	to caulk: 1
to slip off: 2	to pulverize: 2	to gargle: 1
to hear: 2	to remove, extract: 1	to flood: 1
to stretch: 1	to melt, liquefy: 1	to wash, bathe: 1
to hide, conceal: 1	to stretch: 2	to stare: 2
to gape: 5	to look upward: 1	to bare the teeth: 1
to swim: 1	to hit, beat: 1	to break, crack: 1

Display 1 (continued)

to tear, rip: 1	to exceed, pass: 1	to drape over: 1
to wave, flap: 1	to decay, crumble: 2	to attach, join: 1
to break off a piece: 1	to fold: 1	to press, squeeze: 1
to gather: 2	to float: 2	to end, finish: 1
to slice: 1	to tighten: 1	to dive, plunge: 1
to cram, crowd: 1	to sip, suck: 1	to spin: 1
to wash: 2	to budge: 1	to rip, tear, sever: 2
to prune, graze: 1	to pinch: 2	to nod: 1
to snap under tension: 1	to shake, sway: 1	to step, tread: 1
to borrow: 1	to close the eyes: 1	to thrust out, protrude: 1

(B) Nouns

group, company: 1	door, gate: 2	rat, mouse: 1
dust: 2	dew, fog, mist: 3	buttocks: 2
a heap, pile: 1	roof ridge: 1	rice gruel: 1
husk, fibers: 2	mud: 4	ladle, spoon: 1
lid, cover: 2	a bend, curve: 5	tongue: 1
scar: 1	interval: 1	mosquito: 1
lake, pond: 1	cloud: 1	ray of light: 1
fly (insect): 1	rice bug: 1	snout: 1
spitttle: 1	tooth: 1	firefly: 1
dream: 1	bunch, cluster: 1	hook, barb: 2
spot, freckle: 1	rice mortar: 1	suspension bridge: 1
sap: 1	a graceful curve: 1	small protruding part: 1
top, summit: 1	flatulence: 1	wide open space: 1
spider: 1	bag, pouch: 1	

(C) Others

wide: 2	high, tall, above: 2	shallow: 1
rotten (wood): 2	turbid: 1	blind: 1
spotted, striped: 3	dark: 3	pulverized: 1
straight: 1	compressed: 1	small: 2
proud, to boast: 1	dry: 1	confused, lost: 3
angry: 5	stiff, cramps: 3	quick, agile, strong: 1
adhesive, sticky: 1	hunched over: 1	bent, curved: 5
cross-eyed: 1	sweet: 1	sound, of sleep: 1
still, clear (water): 2	tasty: 1	deaf: 1
cruel: 1	flat, level: 3	empty, deflated: 1
thin (materials): 1	dull, blunt: 2	brittle: 1
red: 1	hoarse: 1	ripe, cooked: 1
sour: 1	raw: 1	erect: 1

Display 2: Frequency of association for roots with semantic class/referent type.

bend, curve: 13	shine, glitter: 7	sink, submerge: 7
angry: 5	gape: 5	plug, block, dam: 5
mud: 4	pull out, uproot: 4	scratch, scrape: 4
split: 4	stir, mix: 4	twist, twine, braid: 4
confused, lost: 3	dark: 3	dew, fog, mist: 3
flat, level: 3	insert, penetrate: 3	prop, support: 3
to skin, peel off: 3	shiver, tremble: 3	spotted, striped: 3
tie, wind around: 3	grasp, grip, seize; 3:	splash, squeeze out: 3
stiff, cramps: 3	tie, wind around: 3	bite: 2
buttocks: 2	cover: 2	decay, crumble: 2
door, gate: 2	dull, blunt: 2	dust: 2
fall: 2	float: 2	flow: 2
gather: 2	hear: 2	hook, barb: 2
husk, fibers: 2	high, tall, above: 2	lid, cover: 2
lie prone: 2	open: 2	pinch: 2
pulverize: 2	rip, tear, sever: 2	rotten (wood): 2
shelter, shade: 2	slip off: 2	small: 2
stare: 2	still, clear (water): 2	stretch: 2
to untie, unravel: 2	wash: 2	wide: 2
accompany: 1	adhesive, sticky: 1	to attach, join: 1
bag, pouch: 1	bare the teeth: 1	begin: 1
blind: 1	borrow: 1	break, crack:
break off a piece: 1	brittle: 1	budge: 1
bunch, cluster: 1	caulk: 1	close: 1
close eyes: 1	cloud: 1	compressed: 1
confused, lost: 1	cough: 1	cram, crowd: 1
cross-eyed: 1	cruel: 1	deaf: 1
dive, plunge: 1	drape over: 1	dream: 1
dry: 1	empty, deflate: 1	end, finish: 1
erect: 1	exceed: 1	firefly: 1
flatulence: 1	to flick, spring up: 1	fly (insect): 1
flood: 1	fold: 1	gargle: 1
graceful curve: 1	grope: 1	group, company: 1
grow: 1	hatch: 1	heap, pile: 1
hide, conceal: 1	hit, beat: 1	hoarse: 1
hook, clasp: 1	hunched over: 1	interval: 1
join lengthwise: 1	ladle, spoon: 1	lake, pond: 1
leak, spill out: 1	lean against: 1	look upward: 1
melt, liquefy: 1	mosquito: 1	nod: 1
open the eyes: 1	to press, squeeze: 1	to prune, graze: 1
pulverized: 1	proud: 1	quick, agile, strong: 1

Display 2 (continued)

rat, mouse: 1	raw: 1	ray of light: 1
red: 1	remove, extract: 1	return, restore: 1
rice bug: 1	rice gruel: 1	rice mortar: 1
ripe, cooked: 1	rise, climb: 1	roof ridge: 1
sap: 1	scar: 1	shake, sway: 1
shallow: 1	shrink, shrivel: 1	sink: 1
sip, suck: 1	slice: 1	small: 1
small protruding part: 1	snap under tension: 1	snout: 1
sound, of sleep: 1	sour: 1	spider: 1
spin: 1	spittle: 1	spot, freckle: 1
spread apart, as legs: 1	step, tread: 1	straight: 1
stretch: 1	strew, sow: 1	surround: 1
suspension bridge: 1	swallow: 1	sweet: 1
swim: 1	tasty: 1	tear, rip: 1
thin: 1	think, brood: 1	thrust out, protrude: 1
tighten: 1	tongue: 1	tooth: 1
top, summit: 1	turbid: 1	turn around, revolve: 1
wash, bathe: 1	wave, flap: 1	whip: 1
wide open space: 1	wrap around, as snake: 1	

Appendix 2

Root candidates-in-waiting

The following root candidates are supported by three EIMs, and may eventually qualify as roots if further support can be found.

*-buR	fertile	
1. PWMP	*subuR	fertile, as soil
2. DPB	gabur	fertile (of soil)
3. DPB	gəmbur	fertile (of soil)
*-cek-$_2$	gecko, house lizard	
1. PWMP	*cekcek	gecko, house lizard
2. PWMP	*cikcek	gecko, house lizard
3. Kapampangan	lupísak	house lizard
*-deŋ-$_1$	to extinguish a fire	
1. PAN	*padeŋ	to extinguish, douse a fire
2. PWMP	*pedeŋ	to extinguish, douse a fire
3. PWMP	*pideŋ	to extinguish, douse a fire

(continued)

*-nat	to stretch (dbl. *-ñat)	
1. PWMP	*binat	to stretch
2. Kenyah (Long Wat)	lenat	to strech oneself
3. Kanowit	penat	to strech oneself
*-pir	edge, border (dbl. *-bir)	
1. PWMP	*simpir	edge, margin
2. Amis	ŋapir	edge; ledge from which there is a dropping off place
3. Old Javanese	palipir	border, edge, side

One other root candidate is supported by two EIMs and three forms that are ambiguous for a new root *-cek$_2$ 'to drive in by force, as a post', or the established root *-sek$_2$ 'to insert, stick into a soft surface', as all Philippine forms could reflect either *c or *s:

1. PAN	*pacek	drive in (a nail, post)
2. PPH	*pusek	to press in by force
3. Malay	cacak	planting upright, embedding
4. Cebuano	úgsuk	drive stakes into the ground
5. Cebuano	úsuk	drive stakes into the ground

References

Abuyen, Tomas A. 2000. *Diksyunaryo Waray-Waray [Visaya] English-Tagalog*. Quezon City: Kalayaan Press.
Adriani, N. 1928. *Bare'e-Nederlandsch woordenboek*. Koninklijk Bataviaasch Genootschap van Kunsten en Wetenschappen. Leiden: E.J. Brill.
Alisjahbana, Sutan Takdir. 1949–1950. *Tatabahasa baru Bahasa Indonesia*. 2 vols. Jakarta: P.T. Pustaka Rakjat.
Anceaux, J.C. 1987. *Wolio dictionary (Wolio-English-Indonesian)*. KITLV. Dordrecht: Foris Publications.
Appell, George N., and Laura W.R. Appell. 1961. *A provisional field dictionary of the Rungus Dusun language of North Borneo*. Mimeographed.
Arndt, Paul. 1961. *Wörterbuch der Ngadhasprache*. (Studia Instituti Anthropos 15). Posieux (Switzerland): Paulus Verlag.
Awed, Silin A., Lillian B. Underwood, and Vivian M. van Wynen. 2004. *Tboli-English dictionary*. Manila: Summer Institute of Linguistics.
Ballard, Lee. 2011. *Ibaloy dictionary*. Asheville, North Carolina: Biltmore Press.
Barber, C.C. 1979. *Dictionary of Balinese-English*. 2 vols. (Occasional Publications 2). Aberdeen: Aberdeen University Library.
Barnard, Myra Lou, and Jeannette Forster. 1954. 2nd printing. *Dibabaon-Mandayan vocabulary*. Manila: Summer Institute of Linguistics in cooperation with The Bureau of Public Schools and The Institute of National Language of The Department of Education, Philippines.
Barth, J.P.J. 1910. *Boesangsch-Nederlandsch woordenboek*. Batavia: Government Printing Office.
Beech, Mervyn W.H. 1908. *The Tidong dialects of Borneo*. Oxford: The Clarendon Press.
Behrens, Dietlinde. 2002. *Yakan-English dictionary*. Manila: Linguistic Society of the Philippines.
Bender, Byron W, Ward H. Goodenough, Frederick H. Jackson, Jeffery C. Marck, Kenneth L. Rehg, Ho-Min Sohn, Steven Trussel, and Judith W. Wang. 2003. Proto-Micronesian reconstructions I. *Oceanic Linguistics 42*. 1–110.
Benton, Richard A. 1971. *Pangasinan dictionary*. (PALI Language Texts: Philippines). Honolulu: University of Hawai'i Press.
Berg, René van den. 1996. *Muna-English dictionary*. Leiden: KITLV Press.
Berg, René van den. 2008. *Vitu dictionary*. Computer printout, January 2008.
Bergen, Benjamin K. 2004. The psychological reality of phonaesthemes. *Language* 80. 290–311.
Bergh, J.D. van den. 1953. *Spraakkunst van het Banggais*. KITLV. The Hague: Martinus Nijhoff.
Blust, Robert. 1970. Proto-Austronesian addenda. *Oceanic Linguistics* 9. 104–62.
Blust, Robert. 1974. A Murik vocabulary, with a note on the linguistic position of Murik. *Sarawak Museum Journal* 22.43(NS):153–189.
Blust, Robert. 1976. Dempwolff's reduplicated monosyllables. *Oceanic Linguistics* 15. 107–130.
Blust, Robert. 1982. The Proto-Austronesian word for "female". In Rainer Carle et al., (eds.), *Gava': studies in Austronesian languages and cultures dedicated to Hans Kähler*, 17–30. Berlin: Dietrich Reimer.

Blust, Robert. 1988a. Beyond the morpheme: Austronesian root theory and related matters. In Richard McGinn, (ed.), *Studies in Austronesian linguistics*, 3–90. (Monographs in International Studies, Southeast Asia series 76). Athens: Ohio University Center for International Studies, Center for Southeast Asian Studies.

Blust, Robert. 1988b. *Austronesian root theory: an essay on the limits of morphology.* Amsterdam/Philadelphia: John Benjamins.

Blust, Robert. 1997. Semantic change and the conceptualization of spatial relationships in Austronesian languages. In Gunter Senft, (ed.), *Referring to space: studies in Austronesian and Papuan languages*, 39–51. (Oxford Studies in Anthropological Linguistics). Oxford: Clarendon Press.

Blust, Robert. 2001. Historical morphology and the spirit world: the *qali/kali- prefixes in Austronesian languages. In Joel Bradshaw and Kenneth L. Rehg, (eds.), *Issues in Austronesian morphology: a focusschrift for Byron W. Bender*, 15–73. (PL 519). Canberra: Pacific Linguistics.

Blust, Robert. 2003a. The phonestheme ŋ- in Austronesian languages. *Oceanic Linguistics* 42. 187–212.

Blust, Robert. 2003b. *Thao dictionary.* (Language and Linguistics Monograph Series A5). Taipei: Institute of Linguistics (Preparatory Office), Academia Sinica.

Blust, Robert. 2010. The Greater North Borneo hypothesis. *Oceanic Linguistics* 49. 44–118.

Blust, Robert. 2011. The problem of doubleting in Austronesian languages. *Oceanic Linguistics* 50. 399–457.

Blust, Robert. 2013. *The Austronesian languages.* 2nd rev. ed. (Asia-Pacific Open Access Monographs). Canberra: College of Asia and the Pacific, The Australian National University.

Blust, Robert. 2014. The 'mystery aspirates' in Philippine languages. *Oceanic Linguistics* 57. 221–47.

Blust, Robert. n.d. (1971). Fieldnotes on languages of northern Sarawak, April-November, 1971.

Blust, Robert. n.d.-a. Fieldnotes on languages of central and western Borneo.

Blust, Robert. n.d.-b. Fieldnotes on languages of the Admiralty islands.

Blust, Robert. n.d.-c. Fieldnotes on languages of the Lesser Sunda islands.

Blust, Robert. n.d.-d. Fieldnotes on Formosan languages.

Blust, Robert, and Stephen Trussel. 2020. *Austronesian comparative dictionary.* Online at www.trussel2.com/acd.

Brandstetter, Renward. 1916. Root and word in the Indonesian languages. In C.O. Blagden, *An introduction to Indonesian linguistics, being four essays by Renward Brandstetter, Ph.D.* 1–65. London: The Royal Asiatic Society. [German original published in 1910].

Brewis, Richard, and Keilo A. Brewis. 2004. *Kamus Murut Timugon-Melayu.* Kota Kinabalu, Sabah: Kadazandusun Language Foundation.

Caabay, Marilyn A., Josenita L. Edep, Gail R. Hendrickson, and Melissa S. Melvin. 2014. *Agutaynen-English dictionary, with grammar sketch.* Manila: Linguistic Society of the Philippines.

Capell, Arthur. 1968. 3rd ed. *A new Fijian dictionary.* Suva, Fiji: Government Printer.

Carro, Andrés. 1956 [1888]. *Iloko-English dictionary.* Translated, augmented, and revised by Morice Vanoverbergh, C.I.C.M. Manila.

Cauquelin, Josianne. 2015. *Nanwang Puyuma-English dictionary.* (Language and Linguistics Monograph Series 56). Taipei: Institute of Linguistics, Academia Sinica.

Cense, A.A. 1979. *Makassaars-Nederlands woordenboek.* KITLV. The Hague: Martinus Nijhoff.

Chowning, Ann. n.d.-a. *Bileki (Lakalai, West Nakanai) vocabulary.* Typescript, 284pp.
Chowning, Ann. n.d.-b. *Molima vocabulary.* Typescript, 165pp.
Churchward, C. Maxwell. 1959. *Tongan dictionary.* Tonga: The Government Printing Press.
Clark, Ross. 2009. *Leo Tuai: A comparative lexical study of North and Central Vanuatu languages.* (PL 603). Canberra: Pacific Linguistics.
Codrington, R.H. and J. Palmer. 1896. *A dictionary of the language of Mota, Sugarloaf island, Banks islands.* London: Society for Promoting Christian Knowledge.
Collins, James T. 1979. Expressives in Kedah Malay. In Nguyen Dang Liem, (ed.), *Southeast Asian Linguistic Studies,* vol. 4, 379–406. (PL C-49). Canberra: Pacific Linguistics.
Collins, James T. 2003. *Asilulu-English dictionary.* Jakarta: Badan Penyelenggaraan Seri Nusa, Universitas Katolik Indonesia Atma Jaya.
Collins, Millard A., Virginia R. Collins, and Sulfilix A. Hashim. 2001. *Mapun-English Dictionary.* Manila: Summer Institute of Linguistics, Philippines.
Conklin, Harold C. 1953. *Hanunóo-English vocabulary.* (University of California Publications in Linguistics 9). Berkeley and Los Angeles: University of California Press.
Coolsma, S. 1930 [1884]. 2nd ed. *Soendaneesch-Hollandsch woordenboek.* Leiden: Sijthoff.
Costenoble, H. 1940. *Die Chamorro Sprache.* KITLV. The Hague: Martinus Nijhoff.
Dempwolff, Otto. 1929. *Das austronesische Sprachgut in den polynesischen Sprachen.* Koninklijk Bataviaasch Genootschap van Kunsten en Wetenschappen. Feestbundel I. 62–86.
Dempwolff, Otto. 1938. *Vergleichende Lautlehre des austronesischen Wortschatzes, vol. 3: Austronesisches Wörterverzeichnis.* (Zeitschrift für Eingeborenen-Sprachen Supplement 19). Berlin: Dietrich Reimer.
Devin, Chaumont. n.d. *Buruese-English dictionary.* Typescript.
Diffloth, Gérard. n.d. *Bukat vocabulary.*
Drabbe, P. 1932a. *Woordenboek der Jamdeensche taal.* Bandoeng: Nix.
Drabbe, P. 1932b. *Beknopte spraakkunst en korte woordenlijst der Slaroeëesche taal.* Bandung: A.C. Nix.
Dubois, Carl D. 1976. *Sarangani Manobo: An introductory guide.* Manila: Linguistic Society of the Philippines.
Dunnebier, W. 1951. *Bolaang Mongondowsch-Nederlandsch Woordenboek.* KITLV. The Hague: Martinus Nijhoff.
Eberhand, David M., Simons, Gary F., and Charles D. Fennig, (eds.). 2021. *Ethnologue: Languages of the world,* 24th ed. Dallas, Texas: SIL International.
Elkins, Richard E. 1968. *Manobo-English dictionary.* (Oceanic Linguistics Special Publication 3). Honolulu: University of Hawai'i Press.
English, Leo James. 1986. *Tagalog-English dictionary.* Quezon City: Kalayaan Press.
Esser, S.J. 1964. *De Uma-taal (West-Midden Celebes).* KITLV. The Hague: Martinus Nijhoff.
Ferrell, Raleigh. 1969. *Taiwan aboriginal groups: problems in cultural and linguistic classification.* Institute of Ethnology. (Academia Sinica Monograph 17). Taipei: Academia Sinica.
Ferrell, Raleigh. 1982. *Paiwan dictionary.* (PL C-73). Canberra: Pacific Linguistics.
Fey, Virginia. 1986. *Amis dictionary.* Taipei: The Bible Society.
Firth, J.R. 1930. *Speech.* London: Benn's Sixpenny Library.
Fortgens, J. 1921. *Bijdrage tot de kennis van het Sobojo (Eiland Taliabo, Soela-Groep)* KITLV. The Hague: Martinus Nijhoff.

Forman, Michael L. 1971. *Kapampangan dictionary*. (PALI Language Texts Philippines). Honolulu: University of Hawai'i Press.
Fox, Charles E. 1955. *A dictionary of the Nggela language (Florida, British Solomon Islands)*. Auckland: The Unity Press.
Fox, Charles E. 1970. *Arosi-English dictionary*. (PL C-14). Canberra: Pacific Linguistics.
Fox, Charles E. 1974. *Lau dictionary*. (PL C-25). Canberra: Pacific Linguistics.
Galvin, A.D. 1967. *Kenyah vocabulary*. Privately printed.
Ganang, Ricky, Jay Crain and Vicki Pearson-Rounds. 2006. *Kemaloh Lundayeh-English dictionary*. Computer printout, 421pp.
Geraghty, Paul A. 1990. Review of Blust 1988b. *Anthropos* 85. 530–537.
Goris, R. 1938. *Beknopt Sasaksch-Nederlandsch woordenboek*. Singaraja, Bali: Publicatie Kirtya-Liefrinck-van der Tuuk.
Goudswaard, Nelleke. 2005. *The Begak (Ida'an) language of Sabah*. PhD dissertation, Free University of Amsterdam.
Hapip, Abdul Djebar. 1977. *Kamus Banjar-Indonesia*. Jakarta: Pusat Pembinaan dan Pengembangan Bahasa.
Hardeland, August. 1859. *Dajacksch-Deutsches Wörterbuch*. Amsterdam: Frederik Muller.
Hassan, Irene U., Seymour A. Ashley, and Mary L. Ashley. 1994. *Tausug-English dictionary: Kabtangan Iban Maana*. (Sulu Studies 6). Manila: Summer Institute of Linguistics.
Headland, Thomas N., and Janet D. Headland. 1974. *A Dumagat (Casiguran)-English dictionary*. (PL C-28). Canberra: Pacific Linguistics.
Himmelmann, Nikolaus. 2001. *Sourcebook on Tomini-Tolitoli languages: General information and word lists*. (PL 511). Canberra: Pacific Linguistics.
Hohulin, Richard M., E. Lou Hohulin, and Alberto K. Maddawat. 2018. *Keley-i dictionary and grammar sketch*. Manila: Linguistic Society of the Philippines.
Horne, Elinor Clark. 1974. *Javanese-English dictionary*. New Haven and London: Yale University Press.
Houck, Charlotte, and Nida Quinsay. 1968. *Dictionary of Botolan Sambal*. Ms. held by The Summer Institute of Linguistics, Philippine Branch.
Hudson, Alfred B. 1967. *The Barito isolects of Borneo*. (Data Paper 68). Ithaca: Southeast Asia Program, Department of Asian Studies, Cornell University.
Hughes, Jock. 1995. Dobel vocabulary. In Darrell T. Tryon, (ed.), *Comparative Austronesian dictionary, an introduction to Austronesian studies*, Parts 2–4. Berlin: Mouton de Gruyter.
Ivens, Walter G. 1929. *A dictionary of the language of Sa'a (Mala) and Ulawa, South-East Solomon Islands*. Melbourne: Oxford University Press/Melbourne University Press.
Ivens, Walter G. 1940. *A dictionary of the language of Bugotu, Santa Isabel Island, Solomon islands*. London: Royal Asiatic Society.
Jacobson, Marc. n.d. *Sama Abaknon-English dictionary*. Computer files.
Jaspan, M.A. 1984. *Materials for a Rejang-Indonesian-English dictionary*. (PL D-58). Canberra: Pacific Linguistics.
Jekop, Alica, Benedict Topin, and Rita Lasimbang. 1995. *Kadazan Dusun-Malay-English dictionary*. Kota Kinabalu, Sabah: Kadazan Dusun Cultural Association.
Jonker, J.C.G. 1908. *Rottineesch-Hollandsch woordenboek*. Leiden: Brill.
Josselin de Jong, J.B.P. de. 1947. *Studies in Indonesian culture II: The community of Erai (Wetar)*. (Verhandelingen der Koninklijke Nederlandsche Akademie van Wetenschappen, Afd. Letterkunde). Amsterdam: North Holland Publishing Company.

Kähler, Hans. 1961. *Simalur-Deutsches Wörterbuch mit Deutsche-Simaluresischem Wörterverzeichnis*. (Veröffentlichungen des Seminars für indonesische und Südseesprachen der Universität Hamburg). Berlin: Dietrich Reimer.
Kaufman, Terence. 1990. Review of Blust 1988b. *Language* 66.3. 625–26.
Kawi, Djantera, Abdulrachman Ismail, and Willem Ranrung. 1979–80. *Struktur Bahasa Maanyan*. Jakarta: Pusat Pembinaan dan Pengembangan Bahasa.
Kempler Cohen, E.M. 1999. *Fundamentals of Austronesian roots and etymology*. (PL D-94). Canberra: Pacific Linguistics.
Kimoto, Yukinori. 2017. Documentation and description of the Arta language. (Endangered Languages Archive (ELAR), SOAS), University of London. https://elar.soas.ac.uk.
King, John Wayne and Julie King. 1990. *Sungai/Tombonuwo (Labuk-Sugut)-Bahasa Malaysia-English vocabulary*. (Sabah Museum Publication B2). Kota Kinabalu: Sabah Museum.
King, Victor T. 1976. The Maloh language: a vocabulary and summary of the literature. *The Sarawak Museum Journal* (n.s.) 24 (45).137–164.
Lambrecht, Frans Hubert. 1978. *Ifugaw-English dictionary*. Baguio City, Philippines: The Catholic Vicar Apostolic of The Mountain Province.
Lanyon-Orgill, Peter A. 1962. *A dictionary of the Raluana language*. Victoria, B.C. Privately circulated.
Lewitz, Saveros, and Philip N. Jenner. 1973. Proto-Indonesian and Proto-Mon-Khmer. *Working Papers in Linguistics* 5.10. 113–136. Honolulu: Department of Linguistics, University of Hawai'i Press.
Li, Paul Jen-kuei. 1977. The internal relationships of Rukai. *Bulletin of the Institute of History and Philology, Academia Sinica* 48.1. 1–92.
Li, Paul Jen-kuei. 1978. A comparative vocabulary of Saisiyat dialects. *Bulletin of the Institute of History and Philology, Academia Sinica*: 49.2. 133–199.
Li, Paul Jen-kuei, and Shigeru Tsuchida. 2001. *Pazih dictionary*. (Language and Linguistic Monograph Series A-2). Taipei: Institute of Linguistics (Preparatory Office), Academia Sinica.
Li, Paul Jen-kuei, and Shigeru Tsuchida. 2006. *Kavalan dictionary*. (Language and Linguistics Monograph Series A-19). Taipei: Institute of Linguistics, Academia Sinica.
Lincoln, Peter C. 1977. *Gitua-English vocabulary*. Typescript. 48pp.
Lister-Turner, R., and J. B. Clark. 1930. *A dictionary of the Motu language of Papua*. Sydney: Government Printing Office.
Lobel, Jason William. 2016. *North Borneo sourcebook: Vocabularies and functors*. (PALI Language Texts: Southeast Asia). Honolulu: University of Hawai'i Press.
Maan, G. 1940. *Boelisch-Nederlandsche woordenlijst*. Verhandelingen van het Koninklijk Bataviaasch Genootschap van Kunsten en Wetenschappen. Bandoeng: A.C. Nix.
Macdonald, Charles J-H. 2011. *Palawan-Tagalog-English dictionary* (work in progress). Online resource.
Manik, Tindi Radja. 1977. *Kamus Bahasa Dairi-Pakpak Batak*. Jakarta: Pusat Pembinaan dan Pengembangan Bahasa, Departemen Pendidikan dan Kebudayaan.
Maree, Judith Y.M., and Orlando R. Tomas. 2012. *Ibatan to English dictionary, with English, Filipino, Ilokano, Ivatan indices*. Manila: SIL Philippines.
Marouzeau, J. 1951. *Lexique de la terminologie linguistique français, allemand, anglais, italien*. Paris: Paul Geuthner.
Maxwell, C.N. 1936. Light in the Malay language. *Journal of the Malayan Branch of the Royal Asiatic Society* 14.3. 89–154.

Maxwell, W.E. 1882. *A manual of the Malay language*. London: Kegan Paul, Trench, Trübner, & Co.
McCune, Keith Michael. 1985. *The internal structure of Indonesian roots*, parts 1 and 2. (NUSA, Linguistic Studies of Indonesian and Other Languages of Indonesia 21/22 and 23). Jakarta: Universitas Katolik Indonesia Atma Jaya.
McEwen, J.M. 1970. *Niue dictionary*. Wellington: Department of Maori and Island Affairs.
McFarland, Curtis D. 1977. *Northern Philippine linguistic geography*. (Study of Languages & Cultures of Asia & Africa Monograph Series 9). Tokyo: Institute for the Study of Languages and Cultures of Asia and Africa.
McKaughan, Howard P., and Batua A. Macaraya. 1967. *A Maranao dictionary*. Honolulu: University of Hawai'i Press.
McManus, Edwin G. 1977. *Palauan-English dictionary*. (PALI Language Texts: Micronesia). Honolulu: University of Hawai'i Press.
Mead, David. 1998. *Proto-Bungku-Tolaki: Reconstruction of its phonology and aspects of its morphosyntax*. Ann Arbor: University Microfilms International.
Melalatoa, M.J., Harimurti Kridalaksana, Ahmad Banta, Asiah Abubakar Melalatoa and Abud Daud Benerselam. 1985. *Kamus Bahasa Gayo-Indonesia*. Jakarta: Pusat Pembinaan dan Pengembangan Bahasa.
Meyer, Fr. 1937. *Sikanees-Nederlands woordenboek*. Typescript. 331pp.
Mills, Roger F. 1975. *Proto South Sulawesi and Proto Austronesian phonology*, vol. 1 and 2. Ann Arbor, Michigan: University Microfilms International.
Milner, G.B. 1966. *Samoan dictionary*. London: Oxford University Press.
Mintz, Malcolm Warren, and Jose del Rosario Britanico. 1985. *Bikol-English dictionary*. Quezon City, Philippines: New Day Publishers.
Morris, Cliff. 1984. *Tetun-English dictionary*. (PL C-83). Canberra: Pacific Linguistics.
Morris, Max. 1900. *Die Mentawai-Sprache*. Berlin: Conrad Skopnik.
Motus, Cecile L. 1971. *Hiligaynon dictionary*. (PALI Language Texts: Philippines). Honolulu: University of Hawai'i Press.
Muthalib, Abdul. 1977. *Kamus Bahasa Mandar-Indonesia*. Jakarta: Pusat Pembinaan dan Pengembangan Bahasa.
Nais, William. 1988. *Dayak Bidayuh-English dictionary*. Kuching: Persatuan Kesusasteraan Sarawak.
Naylor, Paz Buenaventura. 1990. Review of Blust 1988b. *Bulletin of the School of Oriental and African Studies* 53.2. 389–90.
Neumann, J.H. 1951. *Karo Bataks-Nederlands woordenboek*. Medan, Sumatra: Varekamp.
Newell, Leonard E. 1993. *Batad Ifugao dictionary, with ethnographic notes*. Manila: Linguistic Society of the Philippines.
Newell, Leonard E. 2006. *Romblomanon dictionary*. Manila: Linguistic Society of the Philippines.
Nickell, Tom, and Kristy Nickell. 1987. *Eastern Agta dictionary*. Summer Institute of Linguistics, Philippines. Computer printout.
Nothofer, Bernd. 1975. *The reconstruction of Proto-Malayo-Javanic*. (Verhandelingen van het Koninklijk Instituut voor Taal-, Land- en Volkenkunde 73). The Hague: Martinus Nijhoff.
Nothofer, Bernd. 1990. Review of Blust 1988b. *Oceanic Linguistics* 22. 132–152.
Nothofer, Bernd. 1991. More on Austronesian radicals (or Roots). *Oceanic Linguistics* 30.
Ogawa, Naoyoshi [with an Introduction by Paul Li]. 2003. *English-Favorlang vocabulary*. Tokyo: Research Institute for Languages and Cultures of Asia and Africa, Asia-African Lexicon Series, no. 43.

Onvlee, L. 1984. *Kamberaas (Oost-Soembaas)-Nederlands Woordenboek*. KITLV. Dordrecht, The Netherlands: Foris Publications.
Pampus, Karl-Heinz. 1999. *Koda Kiwã: Dreisprachiges Wörterbuch des Lamaholot (Dialekt von Lewolema)*. (Deutsche Morgenländische Gesellschaft LII.4). Stuttgart: Franz Steiner.
Pateda, Mansoer. 1977. *Kamus Bahasa Gorontalo-Indonesia*. Jakarta: Pusat Pembinaan dan Pengembangan Bahasa.
Pecoraro, Ferdinand. 1977. *Essai de dictionnaire taroko-français*. (Cahier Archipel 7). Paris: S.E.C.M.I.
Peekel, P. Gerhard. 1929–1930. *Grammatische Gründzuge und Wörterverzeichnis der Label-Sprache*. (Zeitschrift für Eingeborenen-Sprachen 20). 10–34, 92–120.
Pigeaud, Th. 1938. *Javaans-Nederlands handwoordenboek*. Groningen and Batavia: J.B. Wolters.
Post, Ursula, and Mary Jane Gardner. 1992. *Binukid dictionary*. (Studies in Philippine Linguistics 9.2). Manila: Linguistic Society of the Philippines and Summer Institute of Linguistics.
Potet, Jean-Paul G. 1995. Tagalog monosyllabic roots. *Oceanic Linguistics* 34. 345–74.
Pukui, Mary Kawena, and Samuel H. Elbert. 1971. *Hawaiian dictionary*. Honolulu: University of Hawai'i Press.
Rau, D. Victoria, Maa-Neu Dong, and Ann Hui-Huan Chang. 2012. *Yami (Tao) dictionary*. Taipei: National Taiwan University Press.
Reid, Lawrence A. 1976. *Bontok-English dictionary*. (PL C-36). Canberra: Pacific Linguistics.
Reid, Lawrence A., ed. 1971. *Philippine minor languages: word lists and phonologies*. (Oceanic Linguistics Special Publication 8). Honolulu: University of Hawai'i Press.
Reijffert, A. 1956. *Vocabulary of English and Sarawak Land Dyak (Singhi tribe)*. Kuching: Government Printing Office.
Richards, Anthony. 1981. *An Iban-English dictionary*. Oxford: Clarendon Press.
Richardson, J. 1885. *A new Malagasy-English dictionary*. Antananarivo: The London Missionary Society.
Rubino, Carl Ralph Galvez. 2000. *Ilocano dictionary and grammar*. (PALI Language Texts). Honolulu: University of Hawai'i Press.
Ruhlen, Merritt. 1987. *A guide to the world's languages, vol. 1: Classification*. Stanford, California: Stanford University Press.
Sabatier, E. 1971. *Gilbertese-English dictionary* (translated from the original French by Sister Oliva F.N.D.S.C. of the Catholic Mission, Tarawa). Tarawa, Gilbert Islands: Sacred Heart Mission.
Safioedin, Asis. 1977. *Kamus Bahasa Madura-Indonesia*. Jakarta: Pusat Pembinaan dan Pengembangan Bahasa.
Sagart, Laurent. 1994. Proto-Austronesian and Old Chinese evidence for Sino-Austronesian. *Oceanic Linguistics* 33. 271–308.
Said, M. Ide. 1977. *Kamus Bahasa Bugis-Indonesia*. Jakarta: Pusat Pembinaan dan Pengembangan Bahasa.
Schebold, Robert A. 2003. *Central Tagbanwa: A Philippine language on the verge of extinction*. Manila: Linguistic Society of the Philippines.
Schlegel, Stuart A. 1971. *Tiruray-English lexicon*. Berkeley: University of California Press.
Saussure, Ferdinand de. 1959 [1915]. *Course in general linguistics*. Edited by Charles Bally and Albet Sechehaye in collaboration with Albert Riedlinger. Translated with an introduction and notes by Wade Baskin. New York: McGraw-Hill.

Schwarz, J. Alb. T. 1908. *Tontemboansch-Nederlandsch woordenboek*. Leiden: E.J. Brill.
Smith, Alexander D. 2017. The Western Malayo-Polynesian problem. *Oceanic Linguistics* 56. 435–90.
Sneddon, J.N. 1984. *Proto-Sangiric and the Sangiric languages*. (PL B-91). Canberra: Pacific Linguistics.
Southwell, C. Hudson. 1980. *Kayan-English dictionary*. Marudi, Baram, Sarawak.
Steller, K.G.F., and W.E. Aebersold. 1959. *Sangirees-Nederlands woordenboek met Nederlands-Sangirees register*. KITLV. The Hague: Martinus Nijhoff.
Stone, Roger. 2007. *Ayta Abellen dictionary and texts introduction* (work in progress). Online resource available through SIL, Philippines. http://www.philippines.sil.org/resources/works_in_progress/abp.
Stresemann, Erwin. 1927. *Die Lauterscheinungen in den ambonischen Sprachen*. (Zeitschrift für Eingeborenen-Sprachen, Supplement 10). Berlin: Dietrich Reimer.
Strong, Katherine. n.d. *Kanowit-English dictionary*. Computer printout.
Sundermann, H. 1905. *Niassisch-Deutsches Wörterbuch*. Moers: J.W. Spaarmann.
Svelmoe, Gordon, and Thelma Svelmoe. 1990. *Mansaka dictionary*. (LANGUAGE DATA Asia-Pacific Series 16). Dallas: Summer Institute of Linguistics.
Taber, Mark. 1993. Toward a better understanding of the indigenous languages of southwestern Maluku. *Oceanic Linguistics* 32. 389–441.
Tharp, James A., and Maeo C. Natividad. 1976. *Itawis-English wordlist, with English-Itawis finderlist*. (HRAFlex Books, OA29-001, Language and Literature Series). New Haven, Connecticut: Human Relations Area Files Press.
Tim Penyusun Kamus. 1989. *Kamus besar Bahasa Indonesia*. Jakarta: Balai Pustaka.
Topping, Donald M., Pedro M. Ogo, and Bernadita C. Dungca. 1975. *Chamorro-English dictionary*. (PALI Language Texts: Micronesia). Honolulu: University of Hawai'i Press.
Tsuchida, Shigeru. 1980. *Puyuma (Tamalakaw dialect) vocabulary*. Tokyo: Kadokawa Shoten.
Tsuchida, Shigeru, Yukihiro Yamada and Tsunekazu Moriguchi. 1987. *List of selected words of Batanic languages*. Tokyo: Department of Linguistics, Faculty of Letters, University of Tokyo.
Usup, Hunggu Tajudin. 1981. *Rekonstruksi fonem Proto Kelompok Bahasa Gorontalo Sebelah Timur*. Jakarta: Pusat Pembinaan dan Pengembangan Bahasa.
Vanoverbergh, Morice. 1933. *A dictionary of Lepanto Igorot, or Kankanay*. Mödling (St. Gabriel's Abbey, Austria): Anthropos Institute.
Vanoverbergh, Morice. 1972. *Isneg-English vocabulary*. (Oceanic Linguistics Special Publication 11). Honolulu: University of Hawai'i Press.
Veen, H. van der. 1940. *Tae' (Zuid-Toradjasch)-Nederlandsch woordenboek*. KITLV. The Hague: Martinus Nijhoff.
Verheijen, Jilis A.J. 1967–70. *Kamus Manggarai*. 2 parts (Manggarai-Indonesia, Indonesia-Manggarai). KITLV. The Hague: Martinus Nijhoff.
Verheijen, Jilis A.J. 1977. *Bahasa Rembong di Flores Barat, I*. Typescript. Ruteng, Flores, Indonesia: Regio S.V.D.
Verheijen, Jilis A.J. 1982. *Komodo: Het eiland, het volk en de taal*. (Verhandelingen van het Koninklijk Instituut voor Taal-, Land- en Volkenkunde 96). The Hague: Martinus Nijhoff.
Walker, Dale F. 1976. *A grammar of the Lampung language: The Pesisir dialect of Way Lima*. (NUSA 2). Jakarta: Badan Penyelenggara Seri NUSA.
Warneck, Joh. 1977. *Toba-Batak-Deutsches Wörterbuch*. The Hague: Martinus Nijhoff.

Warren, Charles P. 1959. *A vocabulary of the Batak of Palawan*. (Philippine Studies Program, Transcript 7). University of Chicago.
Waterhouse, J.H.L. 1949. *A Roviana and English dictionary*. Revised and enlarged by L.M. Jones. Sydney: Epworth.
Wheeler, Gerald Camden. 1926. *Mono-Alu folklore: Bougainville Strait, Western Solomon Islands*. London: George Routledge & Sons.
Wijngaarden, J.K. 1896. *Savuneesche woordenlijst*. The Hague: Martinus Nijhoff.
Wilkinson, R.J. 1936. Onomatopoeia in Malay. *Journal of the Malayan Branch of the Royal Asiatic Society* 14. 72–88.
Wilkinson, R.J. 1959. *A Malay-English dictionary (romanized)*. 2 parts. London: Macmillan.
Wolfenden, Elmer P. 2001. *A Masbatenyo-English dictionary*. Manila: Linguistic Society of the Philippines.
Wolff, John U. 1972. *A dictionary of Cebuano Visayan*. (Philippine Journal of Linguistics Special Monograph Issue 4). Manila: Linguistic Society of the Philippines.
Wolff, John U. 1974. Proto-Austronesian *r and *d. *Oceanic Linguistics* 13. 77–121.
Wolff, John U. 1982. Proto-Austronesian *c, *z, *g and *T. In Amran Halim, Lois Carrington and S.A. Wurm, (eds.), *Papers from the Third International Conference on Austronesian Linguistics*, vol. 2, 1–30. (PL C 74–77). Canberra: Pacific Linguistics.
Wolff, John U. 1999. The monosyllabic roots of Proto-Austronesian. In Elizabeth Zeitoun and Paul Jen-kuei Li, (eds.), *Selected Papers from the Eighth International Conference on Austronesian Linguistics*, 139–94. (Symposium Series of the Institute of Linguistics (Preparatory Office, Academia Sinica 1)). Taipei: Academia Sinica.
Yamada, Yukihiro. 2002. *Itbayat-English dictionary*. (Endangered Languages of the Pacific Rim A3-006). Kyoto: Nakanishi Printing Co.
Zoetmulder, P.J. 1982. *Old Javanese-English dictionary*. 2 parts. KITLV. The Hague: Martinus Nijhoff.
Zorc, R. David. 1969. *A study of the Aklanon dialect, vol. 2: dictionary*. Kalibo, Aklan: Public Domain.
Zorc, R. David. 1990. The Austronesian monosyllabic root, radical or phonestheme. In Philip Baldi (ed.), *Linguistic change and reconstruction methodology*, 175–94. (Trends in Linguistics Monographs 45). Berlin: Mouton de Gruyter.

www.ingramcontent.com/pod-product-compliance
Lightning Source LLC
Chambersburg PA
CBHW020224170426
43201CB00007B/306